The Law and Regulation of International Space Communication

The Artech House Telecommunication Library

Telecommunications: An Interdisciplinary Text, Leonard Lewin, ed.
Telecommunications in the U.S.: Trends and Policies, Leonard Lewin, ed.
The ITU in a Changing World by George A. Codding, Jr. and Anthony M. Rutkowski
Integrated Services Digital Networks by Anthony M. Rutkowski
The ISDN Workshop: INTUG Proceedings, G. Russell Pipe, ed.
Introduction to Satellite Communication by Bruce R. Elbert
Handbook of Satellite Telecommunications and Broadcasting, L. Ya. Kantor, ed.
Telecommunication Systems by Pierre-Gerard Fontolliet
The Law and Regulation of International Space Communication by Harold M. White, Jr. and Rita Lauria
Introduction to Telephones and Telephone Systems by A. Michael Noll
The Deregulation of International Telecommunications by Ronald S. Eward
The Competition for Markets in International Telecommunications by Ronald S. Eward
Machine Cryptography and Modern Cryptanalysis by Cipher A. Deavours and Louis Kruh
Cryptology Yesterday, Today, and Tomorrow, Cipher A. Deavours et al., eds.
Teleconferencing Technology and Applications by Christine H. Olgren and Lorne A. Parker
The Executive Guide to Video Teleconferencing by Ronald J. Bohm and Lee B. Templeton
World-Traded Services: The Challenge for the Eighties by Raymond J. Krommenacker
The Telecommunications Deregulation Sourcebook, Stuart N. Brotman, ed.
Communications by Manus Egan
Preparing and Delivering Effective Technical Briefings by David L. Adamy
Writing and Managing Winning Technical Proposals by Timothy Whalen
World Atlas of Satellites, Donald M. Jansky, ed.
Communication Satellites in the Geostationary Orbit by Donald M. Jansky and Michel C. Jeruchim
New Directions in Satellite Communications: Challenges for North and South, Heather E. Hudson, ed.
Television Programming across National Boundaries: The EBU and OIRT Experience by Ernest Eugster
Evaluating Telecommunications Technology in Medicine by David Conrath, Earl Dunn, and Christopher Higgins
Optimization of Digital Transmission Systems by K. Tröndle and G. Söder
Optical Fiber Transmission Systems by Siegfried Geckeler
Traffic Flow in Switching Systems by Gerard Hebuterne
Measurements in Optical Fibers and Devices: Theory and Experiments by Giovanni Cancellieri and Umberto Ravaioli
Fiber Optics Communications, Henry F. Taylor, ed.
The Cable Television Technical Handbook by Bobby Harrell

The Law and Regulation of International Space Communication

Rita Lauria White and Harold M. White, Jr.

Artech House
Boston • London

Library of Congress Cataloging-in-Publication Data

Lauria-White, Rita.
The law and regulation of international space communication.

Bibliography: p.
Includes index.
1. Artificial satellites in telecommunication—Law and legislation. 2. Space law. 3. International Telecommunication Union. I. White, Harold M., 1949– II. Title.
K4307.L38 1987 341.4′7 87-30744
ISBN 0-89006-274-9

Cover design adapted from Arthur C. Clarke's view of a three-satellite geosynchronous global communications system

ARTECH HOUSE, INC.
685 Canton Street
Norwood, MA 02062

International Standard Book Number: 0-89006-274-9
Library of Congress Catalog Card Number: 87-30744

10 9 8 7 6 5 4 3 2 1

For each other
and for our children, Camie and Marcela,
and for Mom.

In Memoriam
Wallace J. Lauria, Esq.
Wallace J. Lauria, Jr., Esq.

Contents

Preface

We preface this book briefly, with a suggestion and an explanation. While it is not unusual for a reader only to scan the front matter of a book and not begin reading in earnest until Chapter 1, we nevertheless suggest that the Introduction be read as the beginning of this book. The Introduction has been designed as an integral part of the overall work, and provides fundamental information necessary for the proper understanding of the material that follows.

The explanation relates to the use of sources for the chapters on the history of the International Telecommunication Union and the workings of one of its permanent organs, the International Frequency Registration Board. Because of the antiquity, obscurity, and inaccessibility of original sources in this area, we have drawn heavily for these chapters from secondary sources. We are indeed fortunate, however, that a few impeccable secondary sources exist in this field, sources that are considered definitive works in the area.

One such source is George A. Codding, Jr.'s, *The International Telecommunication Union*,* published in 1952. Codding's classic dissertation is the standard reference on the origins and history of the ITU. Similarly, David M. Leive's *International Telecommunications and International Law: The Regulation of the Radio Spectrum*,** is the standard legal reference on the ITU's regulatory procedures. Leive, General Counsel for INTELSAT, is the recognized expert on legal arrangements for international space communication.

Also drawn from for background information is the collaboration between Codding and former FCC attorney Anthony Rutkowski, *The*

*International Telecommunication in a Changing World.**** Finally, significant insights into the early evolution of space communication law can be gleaned from the classic *Space Law and Government,***** published in 1963 by Andrew G. Haley.

All who work in the field of space communication law and regulation are indebted to these sources. An understanding of the historical events and the evolution of the procedures that govern space communication is indispensable for a comprehensive understanding of the regulatory environment for space communication.

(*) *The International Telecommunication Union.* Leiden, Netherlands: E.J. Brill, 1952.

(**) *International Telecommunication and International Law: The Regulation of the Radio Spectrum.* New York: Oceana Publications, 1970.

(***) *The International Telecommunication Union in a Changing World.* Dedham, MA: Artech House, 1982.

(****) *Space Law and Government.* New York: Appleton-Century-Crofts, 1963.

Acknowledgments

The research for this book was conducted pursuant to a Writing and Research Grant from the National Press Foundation, which supports research and writing projects in order to foster excellence in journalism by advancing the knowledge of the journalism profession. The grant to the authors supported not only the original thesis research on this topic by Ms. Lauria in the School of Journalism at the University of North Carolina at Chapel Hill, but also supported the curricular and research work of Professor White on this topic during a more than two-year visiting professorship in the law, ethics, and policy of space communication and broadcasting in the Department of Radio, Television, and Motion Pictures at UNC-Chapel Hill.

The authors are grateful to many people for the time and help given them during the research, writing, and production of the book. While our thanks go to each person who has helped us in any way during this project, there are a few people whose contributions are so significant that they require special acknowledgment. Quintessential among these is Mr. Steven E. Doyle, Esq., one of the true pioneers of space law. He went far above and beyond the call of duty in closely reading, editing, and commenting extensively upon the entirety of Ms. Lauria's first research draft, the foundation upon which this book is built. Without question, Mr. Doyle's comments were invaluable and, without attributing to him any of our joint errors, omissions, or opinions, we certainly consider him tantamount to an honorary co-author of this effort. Nor would such an accolade be his first. The same is said of his contribution to the 1963 classic *Space Law and Government* by Andrew G. Haley.

We also wish to express our appreciation to Professor George A. Codding of the University of Colorado at Boulder and Professors Jane Delano Brown, Ruth Walden, and Robert Gwyn of the University of North Carolina at Chapel Hill. Each of these individuals read and commented upon one or more of various incarnations of the early drafts of this work. Special mention should be made of Professors Brown and Codding. Without Professor Brown's receptivity, support, and encouragement, the original research for this effort might never have gotten off the ground. Professor Codding is the leading expert on the International Telecommunication Union and kindly agreed to read and comment upon all the material on the origins, history, and evolution of the ITU.

Particular mention must also be made of Professor John Bittner, Chairman of the Department of Radio, Television, and Motion Pictures at UNC-Chapel Hill, Mr. Anthony Rutkowski, Esq., formerly of the Federal Communications Commission, Dr. David Webb, Chairman of the Department of Space Studies at the University of North Dakota, and Ms. Kim Cobrain and the rest of the staff of the International Documents section in Davis Library at UNC-Chapel Hill. The vision of Professor Bittner led to the offering at UNC-CH of pioneering courses in Space Communication Law and the Ethics and Policy of Space Communication. Much of the material herein was tested on UNC students in these courses. Mr. Rutkowski gave generous hours of time for personal consultation during our early search for perspective and primary sources, and Dr. Webb's encouragement and knowledge of original sources in this area, both human and documentary, were a constant source of inspiration. Also, the importance of the generous and enthusiastic support of the Davis Library in acquiring original sources cannot be overestimated.

We are also appreciative of the fine support we have received from the ever helpful professionals at Artech House. Mr. Mark Walsh and Mr. Dennis Ricci saw the potential of the book, established it as a priority on their schedule, and then expedited and assisted in a gentlemanly fashion the production and submission of the final draft. Accolades to Artech House.

Finally, our most special thanks go to our families, whose support, encouragement, patience, and, in the case of our children, significant sacrifice, were indispensable to bringing the project to fruition.

Introduction
The Impact of Space Communication on International Telecommunication Law and Policy

Communication is the process of exchanging information. Communication is necessary for advancement, development, accommodation, and growth — words often used to describe the process of evolution. Communication helps to increase human knowledge and to advance the prospects for cooperation, institutional development, and scientific progress.[1]

Rapid long-distance communication, made possible by international telecommunication, has played an immensely important role in the development of the social, political and economic fabric of modern society. Satellite communication has emerged over the past twenty-five years to become a central component of modern telecommunication systems. Communication via satellite now represents a significant economic activity throughout the world.[2]

The attempt in this book is to examine the complex, often cryptic area of international space communication law and regulation in a manner that can be understood. Designed as a primer — a synoptic reference or handbook that is comprehensible and accessible to the general reader — the book is also intended to prove of value to lawyers, in that it examines and categorizes in detail the complex international regulatory environment for space communication.

Part I begins with a general review of space communication from historical and technical perspectives. We begin in this way because real understanding of the dynamic workings of the international legal environment for space communication requires knowledge of the environmental, technological, and cultural factors that have shaped space communication.

Space communication is only one of the most recent attempts over the ages to achieve rapid communication over long distances. From the broadest historical perspective, its origins can actually be traced back to antiquity, as human attempts to "telecommunicate" far predate the development of any form of electronic communication.

Before the use of electricity, for example, the tradition of communication was mostly oral or written. Towers, or "call-posts," were set up by the ancient Persians where men with loud voices sent messages by shouting from one tower to the next. *Cursus publicus* was an extensive messenger service set up by the Romans, and the House of Taxis operated a form of pony express service thoughout Europe from 1305 to the early 1800s.[3]

Any message, before electricity, had to be manually delivered by people, sent by carrier pigeon, or relayed by semaphore or smoke signal. Each process was slow and tedious. With the harnessing of electricity, however, methods of communication were revolutionized. Telegraphy — electrical communication using wires — helped to usher in the industrial era, and patterns of earthly life were forever changed. Telegraphy allowed for the better management of large, integrated, and far-flung commercial and governmental enterprises.

Similarly, radiocommunication helped to inaugurate the space age and the information age by providing for nearly instantaneous telecommunication with nearly infinite range. This instantaneous communication over long distances made it possible to command and to control rockets and satellites and thereby to achieve light-speed, global communication from high orbits such as the geostationary satellite orbit (GSO).[4]

Although the development of telegraphy preceded that of radio, both technologies were constrained by similar national, political, and social barriers. For example, hostile customs border officials who spoke different languages or used different kinds of receiving equipment diminished the potential for speed, accuracy, and secrecy of the early telegrams. Later, in the early days of radio, a lack of frequency coordination existed and harmful interference was common. The need for coordination became apparent if either telegraphy or radiocommunication was to develop to its maximum potential.[5]

Over the years the realization of this need for coordination and regulation of international telecommunication resulted in a series of international conferences where international treaties were signed regulating telecommunication. These treaties were the forebears of the present International Telecommunication Convention (ITC). This international treaty defines telecommunication as "(a)ny transmission, emission, or reception of signs, signals, writing, images, and sounds or intelligence of any nature by wire, radio, optical or other electromagnetic systems."[6]

Today's worldwide network of international communication required many decades to develop. In Part I we examine the genesis and evolution of the major structural, or institutional, provisions found in the current version of the International Telecommunication Convention. Such provisions create the context for the development of regulatory principles for space communication. That context, the institution for coordination of global telecommunication — the International Telecommunication Union (ITU) — represents a dynamic and ongoing exercise in international cooperation that dates back more than one hundred years.

In Part II we examine the ITU, the primary international institution for the regulation of space communication. The ITU celebrated its hundredth anniversary in Paris at the Salon d'Horloge on 17 May 1965. There on the Quai d'Orsay, the original treaty that established the International Telegraph Union — predecessor to the modern ITU — was signed in 1865.[7]

Within the context of more than a century of international cooperation and technological development, Part II also contains an exploration of the international legal environment for space radiocommunication represented by the structures and functions of the ITU. While telegraph and telephone matters also are handled in the ITU, such matters are covered in Parts I and II only as they relate to the historical evolution of the ITU itself and to the evolution of space communication in particular.

Part III contains an examination of the major ITU plenipotentiary and administrative radio conferences that were concerned with space communication. Policy formulation in the ITU related to space communication began as early as 1959 and has followed a trend established in the early 1960s emphasizing the principle of nondiscriminatory, equitable access to the radio spectrum and the geostationary satellite orbit. The first evidence of this principle is found in ITU Recommendation 10A in 1963.[8] This was followed by Resolution 24 of the 1965 ITU Plenipotentiary Conference, which declared that "all countries should have equal opportunity to use space telecommunication facilities."[9]

Throughout the 1970s, the formulation of space communication policy embraced the principle of nondiscriminatory, equitable access to the radio spectrum and the GSO. This principle was furthered by the adoption of Resolution Spa 2-1 at the ITU space telecommunication conference of 1971. The resolution stated that "all countries have equal rights in the use of both the radio frequencies allocated to various space radiocommunication services and the geostationary satellite orbit for these services."[10]

At the 1973 ITU Plenipotentiary Conference, a version of this equitable access principle was included in the International Telecommunication Convention itself. Article 33 of the 1973 Convention emphasized the efficient and economical use of the radio spectrum and the GSO "so that countries or groups of countries may have equitable access to both . . . according to their needs and the technical facilities at their disposal."[11]

By the end of the 1970s, revolutionary changes had already taken place in the world environment for telecommunication, especially with respect to space communication. At the same time as the demand for communication services was sharply increasing, some portions of the radio spectrum and the GSO were becoming congested. Also by the end of the 1970s, about two-thirds of the membership of the ITU was composed of developing countries, many of which had only recently become independent. This change in the balance of ITU membership resulted in significant changes in ITU policy, especially with regard to space communication.[12]

Basic differences arose between the developed and the developing world in the 1970s with reference to management of the spectrum-orbit resource. Traditionally, the method of securing recognition in the ITU of a legally protected use of the radio spectrum was an *a posteriori*, or first-come, first-served method. The developing countries began to look to an *a priori*, or an advance-planning method of securing legal recognition for satellite communication services to ensure that their future communication needs could be met. Demands were made in the developing world for the principle of equitable access to be put into practice. In contrast, the developed countries saw *a priori* planning as a potentially severe handicap placed upon technological innovation.[13]

With their growing majority, however, the developing countries became more and more able to influence the formulation of law and policy in the ITU.[14] They did so by raising and using the procedure of majority voting rather than relying exclusively upon the traditional approach of reaching general consensus at ITU conferences. The last two chapters of Part III demonstrate that space communication has been the focus of an effort by the developing world, not only to challenge the traditional legal approach to regulating international telecommunication, but also to call for the ITU to develop technical assistance programs to help countries in need of better telecommunication services.

Part IV illustrates how international space communication law and regulation transcends the ITU. This is illustrated through the coverage of a special space communication topic in Chapter 9, a topic that looks at direct broadcast satellites (DBS) in the context of the evolution in the United Nations of principles of international space law. The key unresolved issue of international space law relating to DBS is seen in Chapter 9 to be the issue of the prior consent of receiving states for direct broadcasts into their territories from space.

Satellite communication has become a central issue in the United Nations (UN), with the developing world calling for a New International Information Order. This call for a new information order has been fueled by debate over the transnational flow of information. The debate has concerned the relative merits of a "free" versus a "balanced" or more regulated flow of information across national borders. In this debate the developed democracies have called for freedom of information, while the concern in much of the developing world has been with a "one-way flow" of information from the developed world to the developing world. The concern has been with the effect of this one-way flow on cultural integrity, national development, and national sovereignty.[15]

The developing world has nevertheless seen the use of satellite communication as a way to equalize the flow of information as well as a way to connect their own territories, thus allowing for the origination of information flows from within. As important as satellite communication is for the developed countries, it is perhaps even more important to the developing countries. Many of these countries lack the terrestrial communication infrastructure with which to develop modern communication systems. In the developing world, access to information from satellite communication offers the possibility to vault from a pre-industrial era into the so-called "post-industrial" twenty-first century.[16]

Central to the utility and to the economic advantages of satellite systems is their "connectivity," or ability to establish a mutual connection between a great number of earth stations through a single satellite, thus bypassing the need for a costly terrestrial communication infrastructure, especially into remote areas. Many developing countries believe that the use of satellite communication will thus help them to achieve, in due course, technological parity with the developed world.[17]

We trace in this book the history of international cooperation as it has grown to accommodate the widespread technological capability to use radio and the GSO, resources that transcend national boundaries. Each plateau of technological sophistication and complexity of interaction has led to a new plateau of sophistication in international cooperation. Indeed, several themes can be identified in the developments we recount.

- Change in technology drives social, political, and legal change by driving changes in the nature of commerce, and therefore in social needs and expectations.
- Since technology is based upon knowledge, the currency of this new commerce is information, the flow of which is unavoidably global because of the universality of the radio resource and the broad reach of space communication. This fact increases the interest and dependence of national governments on international law and international communications.
- Satellite communication has thus caused a major reformulation of the law and regulation of international telecommunication. The prime focus of that reformulation is the legal principle of equity — in particular, the principle of equitable access to the geostationary satellite

orbit and the radio frequencies associated with its use.

- This principle of equity has evolved over the past twenty-five years primarily due to the growing influence of the developing countries, which have urged that countries in an earlier stage of evolutionary development should not be foreclosed from equitable access to global resources such as the radio spectrum and the GSO.
- Developed countries, and particularly the United States, have been concerned that truly equitable access not be sacrificed to premature and unwise restrictions on freedom of information or to rigid forms of *a priori* planning of the spectrum-orbit resource. This concern has been expressed not only through tough negotiations and legal compromises but also through the creation of international organizations like the International Telecommunications Satellite Organization (INTELSAT), which was created to help make equitable access to space communication a reality in practice.

With these themes in mind, it should be noted that not all forms of space communication have been swept into the contention concerning equitable access. This is because the generic term "space communication" includes many things such as tracking, telemetry, control, satellite-to-satellite services, crew communications, spacecraft health monitoring, remote-sensing of the earth, communication from deep space missions, and radioastronomy. The contention over equitable access has focused primarily on one area of space communication, that is, satellite communication. Satellite communication essentially is a method of extending the range and capability of terrestrial communication, in that it involves communication originating on Earth, and intended for reception at other points on Earth. What makes such communication "satellite communication" is that the information travels between points on Earth via satellite.[18]

While our coverage throughout the book is illustrative of the international legal environment for all of space communication, it is communication via satellite that has given rise to the prime contention over regulatory approaches. Therefore, even though the ITU recognizes 17 different space services, in this book we have focused on the two most controversial satellite communication services.

The contentious services are the fixed-satellite service (FSS) and the broadcasting-satellite service (BSS). In the FSS, communications are relayed between selected earth stations at specified fixed points. The ITU Radio Regulations legally define the FSS as "a radiocommunications service between earth stations at specified fixed points when one or more satellites are used; in some cases this service includes satellite-to-satellite links, which may also be effected in the inter-satellite service . . . [and] may also include feeder links for other space radiocommunication services."[19]

Since not only telephony but also television is transmitted via the FSS to a virtually unlimited number of such "specified" fixed points, it is obvious that demand for FSS frequencies would be great. The BSS also has served as a substantial impetus for evolving legal principles for space communication. Therefore, of the 17 internationally recognized space communication services, we have examined primarily these two satellite communication services to illustrate the evolving regulation for space communication. A complete list of the 17 space communication services follows.[20]

- Aeronautical Mobile-Satellite
- Amateur-Satellite
- Broadcasting-Satellite (BSS)
- Earth Exploration-Satellite
- Fixed-Satellite (FSS)
- Maritime Mobile-Satellite
- Inter-Satellite
- Meteorological-Satellite
- Mobile-Satellite
- Radionavigation-Satellite
- Space Operation
- Space Research
- Standard Frequency and Time Signal-Satellite
- Aeronautical Radionavigation-Satellite *
- Land Mobile-Satellite *
- Maritime Radionavigation-Satellite *
- Radiodetermination-Satellite *
- (Items followed by asterisk (*) are recognized only in footnotes to the International Radio Regulations. Where not expressly excluded, these services may legally operate in the allocations to the other 13 prime or generic services.)

Again, the contentious issues that have arisen regarding the FSS and the BSS have been the major impetus for evolving legal principles and regulations for space communication in the ITU. Indeed, in 1977, a major break with the traditional regulatory approach in the ITU occurred with respect to the BSS, when the case-by-case, *a posteriori* method of securing recognition of uses of the spectrum-orbit resource was supplanted by *a priori* planning of the 12 GHz band in much of the world.

Yet, this is not to say that other space services will not also be mentioned in this book. Other services such as the earth exploration-satellite service and the mobile-satellite services will be mentioned in passing. For example, the mobile-satellite services also involve communication via satellite and may be destined to play an increasing role in the formulation of space communication law. Already we hear of satellite cellular telephone and satellite ship-to-shore. Moreover, the first radio regulations and service definitions made in the ITU for space endeavors were not even for satellite communication, but rather for space research. Regulations for communication satellites came only later. Still, the primary use of the geostationary orbit has been for point-to-point satellite communication, the primary uses of which (FSS and BSS) are the focus of this book.

Use of the FSS, the most significant satellite communication service in 1987, has evolved considerably over the years. The early earth antennas serving the FSS were large and complex, enabling the satellites to be small and relatively simple. Thus, this type of satellite communication system was the first to be developed. Thus, the FSS became the most heavily used of all the space services and the most highly developed. The large variety of FSS systems handle a wide range of traffic. This includes telephony, telegraphy, television relay, audio conferencing with visual aids, high speed data transmissions, facsimile, communication between computers, and videoconferencing.

Most FSS networks operate in the C band (6/4 GHz) of the radio spectrum. Terrestrial microwave relay systems also operate in this band, sharing it on a primary basis.[21] Of the FSS systems in the Ku band (14/11-12 GHz), most have become operational only since 1980. During the 1980s, studies also have been conducted to determine the parameters of space stations operating in the Ka band (30/20 GHz). Still, the C band is by far the most utilized band (see Figure I.1), not only because the relevant technology is more developed, more widely used, and less expensive in comparison to other systems, but also because the lower frequency bands are less subject to rain attenuation, a factor that is very important to many developing countries located in areas of heavy rainfall.[22]

Finally, among space communication services other than the FSS and the BSS, the number of satellites using the GSO is small, and substantial increases are not foreseen in the near future. (See Figure I.2.) Therefore, as noted above, although the prospects for some of the other space communication services are examined here, the evolving legal principles and regulations for space communication and the arrangements for the FSS and the BSS are the predominant focus of this work. As a prelude to this examination of legal principles, however, we begin with a brief look at the nature of the resources necessary for space communication — the radio spectrum and the geostationary satellite orbit.

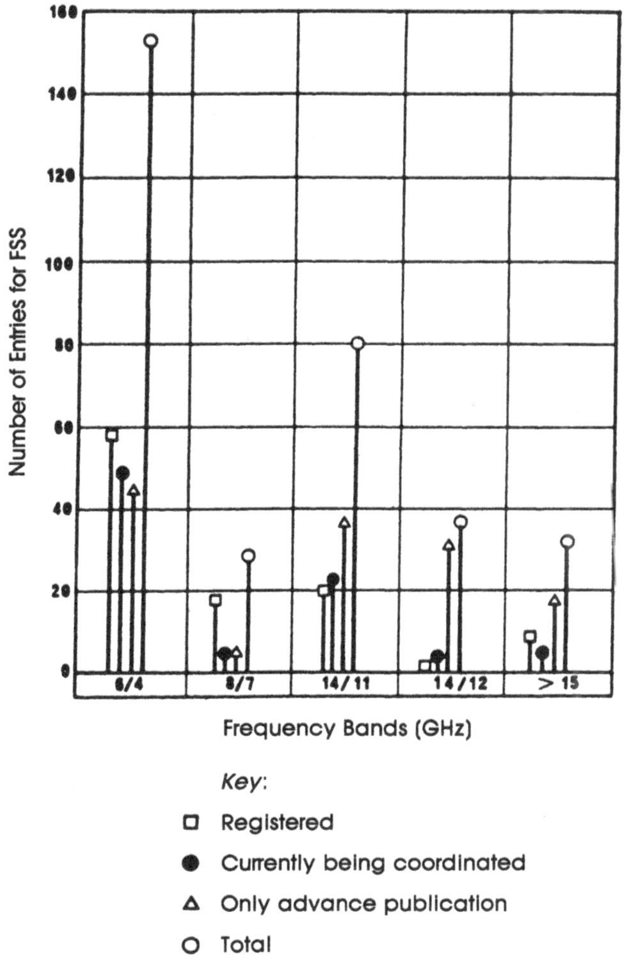

Figure I. 1.A Tabulation of FSS Entries by Frequency Band
Source: Report to the Second Session of the Conference, WARC-ORB-85, ITU, Geneva, 1985.

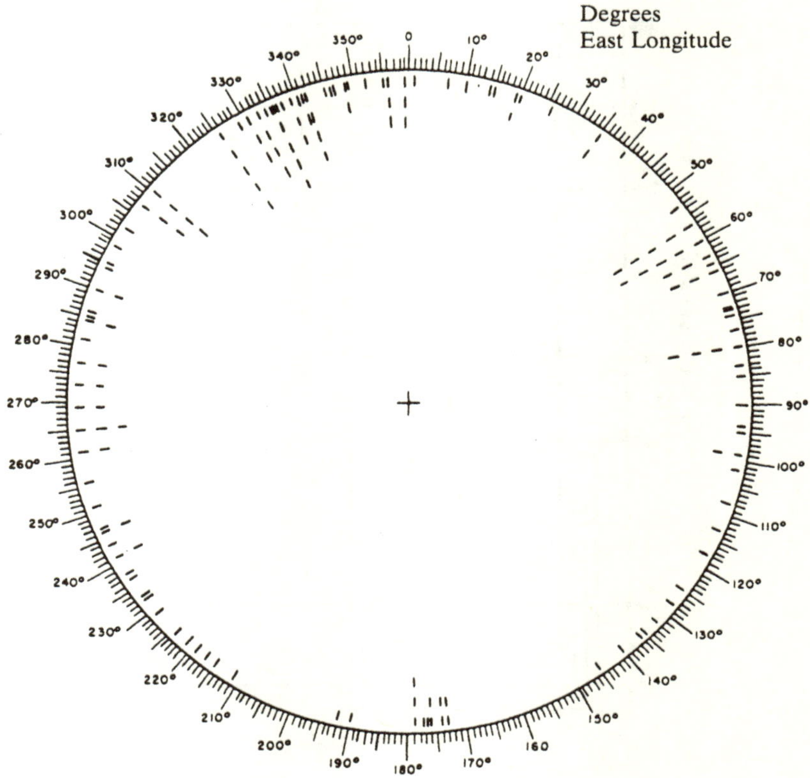

Note: Each dash within the circle represents one space station at a specific orbital position.

Approximate Distribution of Networks
(IFRB data as of Dec. 1983)

FSS only one band	*FSS only ≥ 2 bands*	*Total FSS only*	*FSS and other services*
55%	30%	85%	15%

Figure I. 1.B Orbital Locations of 6/4 GHz FSS Entries
Source: Report to the Second Session of the Conference, WARC-ORB-85, ITU, Geneva, 1985.

International		Domestic	
• Intelsat	~110	• Australia	• Indonesia
• Intersputnik	~15	• Brazil	• Japan
• Inmarsat	~40	• Canada	• Mexico
• Arabsat	~20	• France	• USA
• Eutelsat	~15	• India	• USSR
		• On Intelsat ~ 25	

Figure I. 2 Communication Satellite Systems—An Expanding Population (Courtesy of Stephen E. Doyle, Aerojet Tech Systems Company)

REFERENCES

1. See generally, Pfeiffer, John E., *The Emergence of Man*, New York: Harper & Row, 1969; Pfeiffer, John E., *The Emergence of Society*, New York: McGraw-Hill, 1977; and McLuhan, Marshall, *Understanding Media: The Extensions of Man*, New York: McGraw-Hill, 1964.
2. See generally Martin, James, *Communications Satellite Systems*, Englewood Cliffs, NJ: Prentice-Hall, 1978.
3. Toffler, Alvin, *The Third Wave*, New York: Bantam Books, 1980, p. 32.
4. The geostationary orbit is defined as a circular earth orbit in the plane of the equator at an altitude of about 35,800 km above the Earth's surface. Earth satellites orbit about the center of the Earth with a period determined by the radius of the orbit. At a radius of about 42,165 km, corresponding to an altitude of 35,787 km, a satellite will orbit the Earth in 23 hours and 56 minutes and is therefore synchronous with the Earth's rotation. If the satellite is moving in the same sense as the Earth, that is from West to East, and the satellite orbit is over the equator, the satellite will appear to an observer on Earth to be stationed at a fixed point in the sky, that is geostationary.
5. Feldman, Mildred L. B., *The Role of the United States in the International Telecommunication Union and Pre-ITU Conferences*, Baton Rouge, LA: Mildred L. Bos Feldman, 1976, p. 1.

6. *International Telecommunication Convention, Nairobi, 1982,* (Nairobi Convention), Annex 2, Para. 2015.
7. Codding, George A. Jr., and Anthony M. Rutkowski, *The International Telecommunication Union in a Changing World,* Dedham, MA: Artech House, 1982, p. 44.
8. *Final Acts of the Extraordinary Administrative Radio Conference to Allocate Frequency Bands for Space Radiocommunication Purposes, Geneva, 1963,* Recommendation 10A.
9. *International Telecommunication Convention, Montreux, 1965,* Resolution 24.
10. *Final Acts of the World Administrative Radio Conference for Space Telecommunications, Geneva, 1971,* Resolution Spa 2-1.
11. *International Telecommunication Convention, Málaga-Torremolinos, 1973,* Art. 33.
12. *Radiofrequency Use and Management,* Washington, DC: Office of Technology Assessment, 1981, p. 5.
13. *Ibid.*
14. *Ibid.*
15. See generally Masmoudi, Mustapha, "The New World Information Order," pp. 14-27 and Dordick, Herbert S., "Transborder Data Flows: Competition and Regulation," pp. 210-216 in *World Communications A Handbook,* (G. Gerbner & M. Siefert, eds.), New York: Longman, 1984.
16. Robinson, Glen O., "Regulating International Airwaves: The 1979 WARC," *Virginia Journal of International Law,* Vol. 21, pp. 1-54 at 19.
17. *Radiofrequency Use and Management, supra* note 12, at 75.
18. *ITU Radio Regulations, 1982,* Art. 1.
19. *Ibid.*, Para. 22.
20. *Ibid.*
21. The terms "primary," "permitted," and "secondary" are defined levels of service rights to specified bands. A station of a service designated as secondary in the Table of Frequency Allocations is obligated not to cause harmful interference to stations of a primary or permitted service, irrespective of when the stations of the primary or permitted service started operations or notified the IFRB.
22. World Administrative Radio Conference on the Use of the Geostationary-Satellite Orbit and the Planning of Space Services Utilizing It, First session, Geneva, 1985, *Report to the Second Session of the Conference,* Chap. 2, p. 9.

PART I
BACKGROUND INFORMATION: HISTORICAL AND TECHNICAL

Chapter 1
Radio and the Geostationary Satellite Orbit: The Rationale for Regulation

1.1 INTRODUCTION

Today we live in an exciting time of revolutionary change — change that represents an acceleration of cultural evolution that is bringing unfathomed changes and challenges to this fragile planet.This epoch, already christened the **Information Age**, holds the great potential and promise of propelling the human species to ever higher levels of consciousness, as rapid, perpetual communication — one of the essential keys to higher states of cooperation and understanding — is provided by the proliferation of new telecommunication technologies. Indeed, satellite communication is one of the most striking technological developments devised by human beings to aid our evolving species in its ecological adaptation and in its quest for higher states of consciousness and accomplishment.

All life forms on this planet rely on communication of one form or another for their sustenance and development. The human being is no exception. The central nervous system of the human being is a communication system that operates by almost instantaneously receiving, integrating, and transmitting information. The design and operation of a satellite communication network is an extension and almost perfect template of the central nervous system of the human being. Arthur C. Clarke, in an unpublished speech presented on the occasion of the signing of the Definitive Arrangements for INTELSAT in 1971, referred to the emerging global network of satellite communication systems as the "central nervous system of mankind."[1]

Satellite communication networks — almost instantaneous systems of communication — have played a primary role in pushing our diverse, contemporary world toward the realization of a global village. Satellite communication is a cornerstone of the Information Age and the backbone of the Space Age.

Rapid, global communication in the Information Age, however, relies on the application of a major technological development of the industrial era — radio. Consequently, the International Telecommunication Union treats space communication within the context of space **radio** communication because satellites use radiocommunication for command and control, and for transmission of data or relay of information. Space radiocommunication is defined in the ITU Radio Regulations as "any radiocommunication involving the use of one or more space stations or the use of one or more reflecting satellites or other objects in space."[2]

In this chapter we consider a number of factors that bear upon space communication. First, we examine the nature of radio and the basic principle upon which the regulation of radio is based. Then, we discuss the geostationary satellite orbit (GSO) and some issues related to this prime location in space for communication satellites. Finally, we look at possible technologies and techniques that have been proposed to help ameliorate the problems and issues related to the GSO and the radio spectrum.

The brief discussion that follows is not intended to be exhaustive in technical detail. It merely serves as a primer on technical factors necessary for a basic understanding of satellite communication and the legal issues it presents. The material in this chapter is designed to serve as a foundation and background for our later consideration of the institutional and regulatory environments. The intention is to provide a solid, overall understanding of space radiocommunication so that the reader can more easily understand the macro, evolutionary concepts related to space communication.

1.2 THE NATURE OF RADIO

We begin discussion of the phenomenon of radio by describing it as an arbitrarily defined portion of the electromagnetic spectrum with the dimensions of frequency, time, and physical space. This spectrum is described as "a form of oscillating electrical and magnetic energy capable of traversing space without the benefit of physical interconnections."[3] The rate of oscillation, or frequency, is expressed in cycles per second, or Hertz (Hz), a term named after the German physicist Heinrich Rudolph Hertz who is credited with the discovery in 1886 of radio waves.[4]

The useful electromagnetic spectrum comprises a range of frequencies between 0 and 10^{25} Hz. Within this range are not only the frequencies associated with the most familiar uses of the radio spectrum, such as AM and FM radio and VHF and UHF television, but also the frequencies associated with microwave, infrared, visible light, ultraviolet, x-ray, gamma-ray, and cosmic-ray portions of the electromagnetic spectrum. We are concerned here only with that portion of the electromagnetic spectrum from about 10,000 Hz to 3,000 billion Hz known as the radio spectrum. This range is usually described in terms of 10 kilohertz (kHz) to 3,000 gigahertz (GHz).[5] (See Table 1.1.)

Within this range of almost 3,000 GHz of radio spectrum, only 400 GHz, or about 13 percent of the radio spectrum, have been reduced to regular use in bands allocated by the ITU.[6] Various bands are allocated to particular radio services for use by the nations of the world. This is done by international agreement, with the radio spectrum having been sub-divided into numerous service bands that are allocated on a regional or global basis. Again, the services of primary interest in this work are the fixed-satellite and broadcasting-satellite services (FSS and BSS).

The expansion of the useful radio spectrum to 400 GHz through improved technology occurred in the 1970s. Formal allocation of frequencies occurred in 1979.[7] This allocation was the most recent step in the steady progress of the development of technology to use the radio spectrum occurring since the early part of this century. Previous, successive conferences of the ITU were responsible for allocating ever-expanding radio frequencies for use. In 1912 the radio spectrum was allocated up to 1 MHz, in 1927 up to 23 MHz, in 1938 up to 200 MHz, in 1947 up to 10,500 MHz, up to 40 GHz in 1959,[8] up to 275 GHz in 1971,[9] and finally, up to 400 GHz in 1979.

1.3 RADIO AND THE PRINCIPLE OF HARMFUL INTERFERENCE

The radio spectrum is a natural phenomenon that is uniformly distributed throughout the universe. It would appear, in theory, that the radio spectrum is infinite. This phenomenon has been characterized not as infinite, however, but as a "limited natural resource." One reason for this characterization is that technology has not yet advanced to the point where humanity can use the entire radio spectrum. Another reason is that the use of radio can be limited by interference, and thus must be skillfully and cooperatively managed.

The natural limitations on the electromagnetic spectrum include attenuation due to the constituent gases in the atmosphere and absorption by water vapor and other constituents such as clouds, fog, or precipitation. The man-made limitations on the use of the radio resource are due pri-

Table 1.1 Typical Uses of Spectrum by Major ITU Band

Source: Levin, Harvey J., *The Invisible Resource: Use and Regulation of the Radio Spectrum,* Baltimore: Johns Hopkins University Press. Published for Resources for the Future, Inc. Reprinted with permission.

Frequencies	*Band Number*	*Band*	*Typical Uses*
300–3000 GHz*			Unallocated radio spectrum.
30–300 GHz	11	Extremely High Frequencies (EHF)	Microwave relay; space research; radar; radionavigation; amateur, experimental; radio astronomy.
3–30 GHz	10	Super High Frequencies (SHF)	Microwave relay; deep space, space research, telemetry, communications satellites; radar; aeronautical radionavigation; meterological aids; amateur, citizens; radio astronomy.
300–3000 MHz	9	Ultra High Frequencies (UHF)	Short-range communications; microwave relay; over-the-horizon "scatter" communication. UHF television, instructional TV; land mobile; weather satellites, meteorological aids; space tracking and telemetry; radar; worldwide aeronautical radionavigation; amateur, citizens, radio astronomy.

Table 1.1 (cont'd)

Frequencies	*Band Number*	*Band*	*Typical Uses*
30–300 MHz	8	Very High Frequencies (VHF)	Short-range line-of-sight communication; over-the-horizon "scatter" communication. VHF television, FM broadcasting; space tracking and telemetry, satellites; aeronautical distress; worldwide radionavigation; land mobile; amateur; radio astronomy.
3–30 MHz	7	High Frequencies (HF)	Medium- and long-range communication. International broadcasting; international point-point; air-ground; ship-shore; space research; amateur, citizens, radio astronomy.
300–3000 kHz	6	Medium Frequencies (MF)	Medium- and short-range communication. AM broadcasting; aeronautical mobile, radionavigation, marine radiophone, Loran; international distress, disaster; amateur.
30–300 kHz	5	Low Frequencies (LF)	Long- and medium-range point-point communications; radionavigation aids; aeronautical mobile.
10–30 kHz	4	Very Low Frequencies (VLF)	Very long-range point-point communications (over 1000 nautical miles).

*The radio spectrum was allocated up to 400 GHz by the ITU in 1979.

marily to interference between radio systems that simultaneously use the same radio spectrum in the same area. Interference can result from signals intended for a different service area or from signals intended for a different purpose.

The concept of harmful interference involves the degradation of the performance of a radiocommunication system due to unwanted radio signals from other systems. The concept is specifically described in the International Telecommunication Convention (ITU Convention) as "interference which endangers the functioning of a radionavigation service or of other safety services or seriously degrades, obstructs or repeatedly interrupts a radiocommunication service operating in accordance with the Regulations."[10]

Indeed, two fundamental reasons for the existence of the International Telecommunication Union are:

- to "effect allocation of the radio frequency spectrum and registration of radio frequency assignments in order to avoid harmful interference between radio stations of different countries;" and
- to "coordinate efforts to eliminate harmful interference between radio stations of different countries. . . ."[11]

Radio interference is caused in various ways. Signals transmitted on a given frequency occupy all three spectral dimensions of time, space, and frequency. Two radio operators cannot use the same radio frequency at the same time in the same geographic area. If two or more users choose to operate on the same frequency at the same time, interference will result unless they are far enough apart geographically, or the transmitted signal is so configured (polarized) that they will not interfere. Therefore, "guard bands" have been established between active bands to help reduce the possibility of accidental interference through overlap of adjacent frequencies.[12]

These are only some of the ways in which harmful interference can result unless the world's radio users undertake strict coordination and compliance with regulations for the use of radio. The rules established by the ITU for radio use are basic, derive from the scientific nature of radio and the technological nature of telecommunication, and constitute natural limitations on freedom of use of the radio spectrum. Simply put, if the technological capability for radiocommunication is to be successfully engaged, cooperation is required. Thus, change in technology is driving social, political, and legal change.

1.4 THE GEOSTATIONARY ORBIT

In his famous *Wireless World* article of 1945, entitled "Extraterrestrial Relays," Arthur C. Clarke suggested that "a true broadcast service giving constant field strength at all times over the whole of the globe would be invaluable, not to say indispensable, in a world society." He proposed that this could be accomplished by three strategically spaced satellites in the geostationary orbit, a circular orbit above the equator at an altitude of about 35,800 kilometers (22,300 miles). The geostationary orbit is unique in that the time it takes for a satellite to complete this orbit of 265,000 kilometers is 23 hours and 56 minutes, approximately one day.[13]

Satellites in the GSO are, therefore, synchronous with the earth's rotation. To an observer on earth, such satellites will appear to be at a fixed position in the sky, i.e., geostationary. As Clarke revealed, a geostationary satellite has a constant view of about 40 percent of the earth's surface and is continuously visible from any point within that area, therefore allowing for continuous transmission (see Figure 1.1). Fixed ground antennas, consequently, do not need to be continually reorientated to track the satellite. Hence, the cost of computer-controlled tracking equipment is avoided. Only three such geostationary satellites are necessary to cover nearly the entire surface of the Earth.

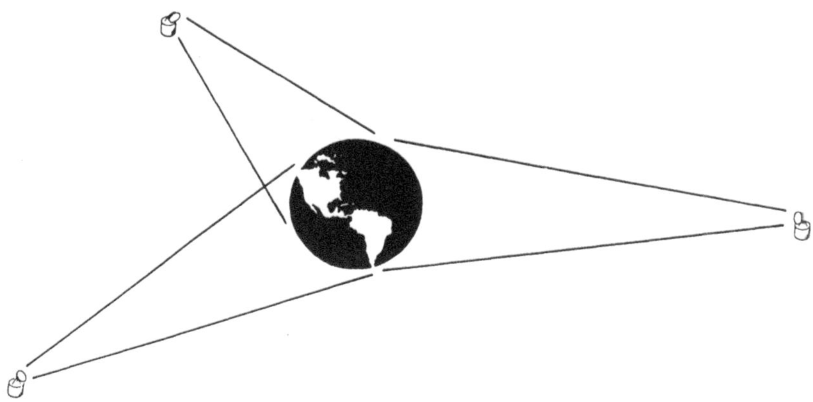

Figure 1.1 Three Geostationary Satellites Can Cover the Populated Portions of Earth
(Courtesy of Stephen E. Doyle, Aerojet Tech Systems Company)

In addition to satellite communication, the GSO is used for other satellite systems like meteorological-satellite, space research-satellite, and earth research-satellite services, among others. These others include, for example, the US tracking and data relay satellite system (TDRS), which can track satellites in lower orbits and relay data to and from many such satellites with the use of a single ground station. The TDRS is a dramatic example of the GSO's special advantages. Only two of these geostationary satellites are replacing the US satellite tracking data network (STDN), a worldwide overseas network of expensive, complex, continually staffed, tracking ground stations.[14]

1.5 THE ISSUE OF ORBITAL CROWDING

Since the first geostationary satellite was launched in 1963, the number of satellites in the geostationary orbit has continued to grow. Over the last decade, the number of satellites in this orbit increased about 18 percent per year.[15] Like the radio spectrum, the GSO is widely considered a limited natural resource. Although this orbit can theoretically hold a multitude of satellites without a high risk of collision, the number of satellites that can operate in the GSO on a given frequency without consequent radio interference is much more limited. The issue of crowding in the GSO is a highly politically-charged debate, with many divergent viewpoints.

The lesser developed countries, in particular, hold that while they are becoming technologically and economically able to place satellites in the GSO, orbital positions for their satellites are rapidly diminishing. The more developed nations maintain that technological advances will preclude any real congestion of the orbit. Thus, the developed nations feel that a more laissez-faire approach to regulation of the GSO is in order, while the developing nations call for strict planning procedures and regulations.

According to a 1985 advisory report to the US Federal Communications Commission (FCC), by the end of 1984 there was a total of 138 satellites of all types operating in the GSO. These included 80 communication satellites, about 48 of which had been launched since 1979. About 14 of those launched during this period serve the developing countries (see Table 1.2). Of the total 80 communication satellites, 55 operate exclusively in the C band (6/4 GHz) and 15 operate exclusively in the Ku band (14/11-12 GHz), with the remaining 10 satellites being hybrid satellites using both bands. The report also noted that about 160 additional communication satellites, including replacements, had been proposed for launch within five years.[16] With improved technology, the most common design lifetimes for communication satellites launched during the mid-1980s is ten years.[17]

Table 1.2 Number of Operational Communication Satellites in Geostationary Orbit*
Source: FCC, Second Advisory Committee Report for the ITU WARC-ORB, Washington, DC, January 1985.

	6/4 GHz	*14/11-12 GHz*	*Hybrid*	*Total*
North/South America (30° W—180° W)	16	11	5	32
Europe/Africa/Mid-East (30° W—66° E)	17	3	4	24
Asia/Far East (67° W—180° E)	22	1	1	24
Subtotal	55	15	10	80

*The numbers in the table reflect communications satellites in GSO, derived from the IFRB's Master Satellite (International Frequency) Register and other sources. There were a total of 138 satellites of all types confirmed to be in the GSO, as of 1984.

This growing number of satellites in service creates a difficult situation for countries trying to find orbital positions within specific segments of the orbital arc that can service their particular area. It also makes it difficult to secure preferred frequency bands. Indeed, there are certain orbital segments and frequency bands that are already overcrowded.

Two factors limit the number of satellites that can use the geostationary orbit — physical interference between satellites and radio frequency interference between systems. Here the limit to the capacity of the orbit will be expressed in terms of the number of satellites subject to collision and radio frequency bandwidth subject to interference.

1.6 COLLISION PROBABILITIES IN THE GSO

Due to orbital variations resulting from natural factors like the ellipticity of the earth's equator, the gravitational attraction of the sun and moon, and solar radiation, a satellite will tend to drift in longitude, moving north and south of the equator in a figure-eight pattern. To counter this drift, satellites are designed with *station-keeping* systems on board. These systems keep the satellite in the desired position within the orbit. Current geostationary satellites can maintain a position to an accuracy of plus or minus 0.1 degree, both in longitude and in latitude. This corresponds to an area of space 150 kilometers from north to south and from east to west. The satellite also will vary in altitude by about 30 kilometers. The GSO, therefore, can be looked at as a ring or torus in space about 150 kilometers from north to south and 30 kilometers thick.[18]

Since most contemporary satellites can maintain their position within plus or minus 0.1 degree of longitude, a minimum of 0.2 degree of separation is required between nominal orbital positions to avoid collision. A nominal orbital position is the longitudinal position in the GSO associated with a frequency assignment to a satellite. Thus, at current station-keeping tolerances, there are 1800 potential orbital positions available in the GSO. Even at this minimum spacing there is little theoretical risk of collision between functioning satellites. In practice, however, satellites actually are positioned at least two degrees or more apart. At two degrees spacing, or about 1500 kilometers apart, the probability of collision is extremely low. Even if ten or more satellites are placed in the same nominal orbital position, the risk of collision is very small. In a 1980 report it was estimated that if ten satellites, each with a 100 square meters cross sectional area, were placed in the same nominal position, the probability of collision would be only 4×10^{-5} per year (0.00004) — that is an average of one collision between active satellites every 400,000 years.[19]

The main risk of collision to active satellites in the GSO results from inactive satellites drifting into the orbit. The same 1980 report estimated the probability of collision between active and inactive satellites to be relatively small — 9.7×10^{-4} (0.00097) per year — based on an assumption that 110 active and 310 inactive satellites are in the GSO. The risk of collision begins to increase substantially only when large systems like solar power satellites (SPSs) with large solar panel arrays are placed in the orbit. Then the probability of a hit increases. Assuming four SPSs in the GSO, each with an average area of 50 square kilometers, in addition to 120 active and 310 inactive satellites, the risk of collision becomes one every five years.[20] (See Table 1.3.)

Table 1.3 Collision Probabilities in the GSO
Source: "Efficient Use of the Geostationary Orbit" UN Doc. A/CONF. 101/BP/7, 1981.

	Satellites		*Average*		
Period	*Active*	*Inactive*	*Area*	*Probability*	*Notes*
1981–1985	95	65	13 m^2	0.3×10^{-4}	
1986–1990	135	140	21 m^2	1.2×10^{-4}	
1991–1995	115	250	124 m^2	5.5×10^{-4}	
1996–2000	110	310	168 m^2	9.7×10^{-4}	no SPS
1996–2000	120	310	1.7 km^2	0.98	4 SPS

Since the major risk of collision between satellites is from drifting, inactive satellites, it has become regular practice to remove satellites from the GSO at the end of their useful lifetimes. It has been suggested that since inactive satellites appear to drift in altitude by less than 100 kilometers, they could be raised into stable circular orbits 100 to 200 kilometers above the GSO. This would eliminate the risk of collision. It should be noted that the failure of the station-keeping propulsion system before the planned end-of-lifetime of the satellite would prevent removal, but this type of failure is relatively rare.[21]

Even though it would appear that there are many unused orbital positions and that the likelihood of collision is low, a few important considerations need to be acknowledged when discussing congestion of the GSO. These considerations involve demand for the most suitable orbit location for a particular system, using economic and reliable technologies, and demands for the most widely used frequencies, namely the 6/4 GHz bands. These considerations are of primary concern to the developing countries. In order to understand demands for positions in particular orbital segments, it is useful here to introduce a few definitions.

1.7 AREA OF VISIBILITY, SERVICE AREA, AND COVERAGE AREA

The area of the Earth's surface that is visible from a geostationary satellite is a circle with a radius of 9,050 kilometers about the subsatellite point on the equator. This extends in latitude from roughly 81.3 degrees north to 81.3 degrees south and in longitude from 81.3 degrees east to 81.3 degrees west of the subsatellite point. The satellite is visible from every point in this circle, appearing at the zenith from the center and on the horizon from the circumference. Because of atmospheric attenuation, however, a satellite must be above the horizon for reliable communications. Without powerful, INTELSAT–class earth stations, the minimum useful elevation is 10 degrees, even greater in mountainous areas. This corresponds to a circle with a radius of about 7,952 kilometers (see Figure 1.2). This area is called the **area of visibility** and has a width that depends on the latitude of the satellite. At 70 degrees north or south latitude, the area of visibility is only 1,640 kilometers, while at 71.43 degrees it is zero.[22]

The **service area** of a satellite is the area in which a useful transmitted signal can be received by an earth station. If the width of the service area equals the width of the area of visibility, then the satellite can only be located at the central longitude of the service area and thus has a very short service arc. If the service area is appreciably smaller than the area of visibility, there are more options for the satellite's location, with the

Latitude	*Area*
0° (equator)	15,900 km
30° N or S	13,200 km
45°	10,000 km
60°	5,620 km
70°	1,640 km
71,43°	0 km

Figure 1.2 Geostationary Satellite Area of Visibility by Latitude
Source: "Efficient Use of the Geostationary Orbit," UN Doc. A/CONF. 101/BP/7, 1981.

satellite needing only to be located within a range of longitudes about the central point. Consequently, those satellites serving large areas, like global satellite communication systems, need specific positions in the geostationary orbit. On the other hand, those satellites serving smaller areas can be positioned more flexibly. Finally, the **coverage area** of a satellite is the area effectively covered by the communication antennas of the satellite.[23]

These constraints on satellite positioning as well as limitations on the radio spectrum are the primary reasons the GSO has been described as congested. It should be pointed out, however, that the GSO is not uniformly populated, nor is it likely to be so in the future. Still, those portions of the orbit that serve large areas of heavy communication traffic are virtually full with respect to the typical communication satellite — one that operates in the C band with unpolarized transmissions covering the entire area of visibility.

The more populous portions of the geostationary orbit include the areas of the geostationary arc serving North America from about 135 degrees west to 87 degrees west longitude, the area over the Atlantic Ocean serving Western Europe from about 1 degree west to 35 degrees west longitude and the area over the Indian Ocean serving Eastern Europe and the USSR from about 49 degrees east to 90 degrees east longitude. Growth in traffic using the C band in these areas can be accommodated only by using various spectrum-orbit utilization enhancement techniques or by switching to other bands.[24]

1.8 FREQUENCY CONGESTION

Limitations on the use of the radio frequency spectrum may pose the greatest problem to increased use of the GSO. Prior to the 1980s, most communication satellites used frequencies below 10 GHz. At frequencies above 10 GHz, attenuation of the signal occurs in heavy rainfall, and at higher frequencies the constituent gases of the atmosphere pose a greater problem.

Information that is transmitted by radiocommunication occupies a certain range of frequencies, or bandwidth. The bandwidth depends on the amount of information to be carried and on the way the radio waves are modulated to carry the information. To avoid interference, which cannot be eliminated entirely, the geographic areas and the frequency range occupied by a communication satellite include a minimum service area, a minimum bandwidth for the earth station and the satellite to transmit their signals, and a bordering geographic area and frequency bandwidth over which the transmission is reduced to a level that will not interfere with other systems.[25]

The transmissions up to the satellite (up-link) and back down to the earth stations (down-link) are at separate frequencies to avoid interference between the energy being sent and that being received. Between geostationary satellites, interference also is avoided by separating the satellites in space as well as in frequency. Typically, satellites operating in the 6/4 GHz bands are spaced between three and five degrees apart, although spacing of two degrees has recently been introduced because of improvement in satellite antenna design.[26]

For example, by the mid-1980s, because of higher performance electronics and antenna designs, many systems in the United States could achieve a good telephone channel in only 3 kHz of bandwidth. In 1980, by contrast, the global norm for a telephone or voice channel on a typical, or standard, communication satellite occupied about 4 kHz of bandwidth, while a television channel occupied a bandwidth of about 40 MHz. A typical 40 MHz satellite transponder, at this level of technical sophistication, carried either 1,000 voice channels or one television channel. An example of a commonly used frequency bandwidth for a communication satellite transmission to a ground station is 3.4 GHz to 3.9 GHz — a band 500 MHz, or 0.5 GHz wide. A typical satellite might have 12 transponders of 40 MHz bandwidth each within this band. This is a total of 12,000 voice channels or 12 television channels.[27]

Again, until recently, the lower frequency C band was most in demand because the technology for this frequency range was the oldest, the cheapest to operate, and the least susceptible to rain attenuation. The keen competition for use of the C band, frequencies that have been widely used and shared by both terrestrial and space communication services, established the historical basis for claims of congestion of the GSO and radio spectrum. Since 1980, however, expansion of services into higher frequency ranges has increased production and lowered costs for the Ka band (14/

11-12 GHz) and substantial improvements in system designs and operational technologies have reduced the limitations of atmospheric attenuation, leading many of the developed nations of the world to maintain that technological innovation can solve the problems related to congestion.

1.9 TECHNOLOGIES AND TECHNIQUES FOR MORE EFFECTIVE USE OF THE GSO

1.9.1 Signal Polarization

Again we see the inexorable development of technology underlying political and legal accommodation when we examine some of the numerous techniques that are used to increase the amount of information transmitted in a given bandwidth. Orthogonal polarization, for example, is a frequency reuse technique that enables two signals to be sent and received independently at the same frequency. In the linear mode, the electric and magnetic fields are orientated in fixed planes, allowing both vertical (north-south) and horizontal (east-west) polarization of the electromagnetic radiation. This effectively doubles the amount of information that can be transmitted in a given bandwidth. There is also circular polarization where the electromagnetic energy is sent in opposite directions. A number of satellites currently in orbit, like Intelsat V, use a dual polarization technique to get an effective 1,000 MHz bandwidth in a nominal band of 500 MHz bandwidth.[28]

A technical problem related to frequency reuse techniques, however, is depolarization of the signal due to precipitation. Development efforts are demonstrating that it is possible to compensate for the depolarized signal at the earth station by employing a beam that can be read at alternative earth station sites outside of the heavy precipitation area.[29]

1.9.2 Reverse Bandworking

Reverse bandworking is another way to provide a significant degree of resource enhancement for the FSS. This technique uses the same frequency band for up-links to one satellite as for down-links from another adjacent satellite. Current practice is to assign different frequencies for the earth station to satellite transmission than for the satellite to earth station transmission. In reverse bandworking, up-link and down-link frequencies are thus alternated between adjacent satellites, so that, for example, a satellite using a 4/6 GHz transmission pattern might be placed between two satellites using a 6/4 GHz transmission pattern.

Use of this technique is compatible with current technologies and could possibly increase the number of satellites that could use a given portion of the orbit by 70 to 90 percent. However, its implementation would add some complexity to coordination with existing systems sharing the same frequency band and would necessitate some means to deal with interference to and from terrestrial services in those bands with shared use.[30]

1.9.3 Advanced Antenna Design

Use of advanced antenna technologies also can help. Antennas used for satellite communication, both on the ground and on the satellite, have a beamwidth, or field of view, that depends upon the frequency used and upon the diameter of the antenna. The size of the beamwidth is a function of the antenna size and the frequency used, decreasing as the antenna diameter and the frequency increase. A satellite antenna can have a very wide beam to cover a large area of visibility, or it can have a narrowly shaped, or **spot beam** to cover a small geographic area. Antenna feed horns are used to shape the beam. Use of overlapped "designer" spot beams can be made to cover a specific area.

The use of separate beams for satellite transmissions is another frequency reuse technique. It allows the same frequency to be used independently in each beam, thus increasing the system capacity by a factor equal to the number of beams. For example, if a satellite providing North American domestic service uses directional antennas, it can effectively avoid interfering with a South American service using the same frequencies via a satellite in the same general area of the GSO.

Beam separation is primarily advantageous when demand is widely distributed geographically. It cannot increase capacity within a small region of high traffic density. The Intelsat V satellite, which services a global area, is effectively using this technique to get a fourfold increase in capacity. It does this by using beam separation to double the capacity of the 6/4 GHz band, in addition to the doubled capacity brought about by dual polarization. Similar separate beam use is being implemented for coverage of noncontiguous geographical areas and time zones covered by domestic US satellites.[31]

At any given frequency, however, the minimum spacing between satellites is defined not only by the transmission beamwidth but also by the level of radiation in the antenna sidelobes. Antennas, rather than radiating energy only into the transmission beam, actually emit significant energy in other directions, particularly into the antenna sidelobes adjacent to the main beam. This level of radiation can interfere with communications. Consequently, the spacing distance of the satellite is determined in part by the power levels in the sidelobes and the sensitivity of adjacent systems to interference. Advanced antenna designs are being developed to reduce sidelobe radiation and allow for closer spacing between satellites. Reducing satellite system sensitivity to sidelobe interference could also permit closer spacing.[32]

The situation in the broadcasting-satellite service (BSS) is different because the receiving antennas on earth are relatively small and inexpensive in contrast to those used in the FSS. Direct broadcasting satellites (DBS), not yet widely implemented, operate on bands allocated for the BSS. DBS systems generally are characterized by high-power satellite signal transmissions capable of being received by low cost equipment with a small (one-meter diameter or less) antenna and a simple receiver-converter. Such systems can be designed for direct broadcasting of television to community or to individual home receivers. Consequently, the satellite, which must transmit high-power signals, can cause problems of interference between satellites in this and other services.

Solutions to the power-level problem include reducing the satellite's transmitted power by improving the performance and sensitivity of the receiving system. This needs to be done without diminishing the basic concept underlying the BSS — that it is intended for direct reception by the general public with small and relatively low-cost receiving stations.

1.9.4 Orbit Sectorization

Inhomogeneity between satellite systems also can result in less effective orbit utilization, the potential for which cannot be completely eliminated because of the wide variety of applications for which the FSS is used. One of the biggest opportunities to prevent ineffective use of the orbit is to coordinate satellite systems development during the design phase. Another alternative is to use orbit sectorization or spectrum segmentation to lessen the impact of highly incompatible emissions.[33]

Orbit sectorization involves positioning systems with similiar characteristics (homogeneous systems) in specific sectors of the GSO. It may be desirable for satellites that have different antenna diameters, receiving sensitivities, *et cetera* to be located in different sectors of the orbit. Although orbit sectorization could reduce the problem of inhomogeneity without constraining system characteristics, it most likely would impose constraints on the choice of orbital positions for satellites. Since networks serving large service areas or high latitudes have small service arcs, severe problems could arise for some countries seeking orbital positions for their broad-coverage satellite systems.

Thus, there may be a need for **guard arcs** to avoid severe inhomogeneity at the interface between sectors. Orbit sectorization could provide large benefits, especially when the services to a specific geographic area are harmonious or where there are regional differences in frequency allocations. The first session of the World Administrative Radio Conference on the Use of the Geostationary-Satellite Orbit and the Planning of Space Services Utilizing It (WARC-ORB) that was held in 1985 called for further study to assess more accurately the advantages and disadvantages of orbit sectorization.[34]

1.9.5 Spectrum Segmentation

Spectrum segmentation is another approach to using the GSO more effectively. It is based on the fact that high spectral density carriers like television cause high interference to low-capacity radio carriers like single channel per carrier (SCPC) carriers, which might carry a single telephone channel. Spectrum segmentation would reduce inhomogeneity within systems by segregating the frequencies used by high-density and low-capacity carriers.

The use of the same frequency by carriers of differing capacity currently necessitates large spacing distances between satellites. It also has been suggested to segment frequency bands by defining the two edges of a band and then placing the high density carriers like TV at one edge of the band and SCPC carriers at the other edge. Again, further study has been called for by WARC-ORB-85 to assess how the potential benefits of spectrum segmentation can be achieved.[35]

1.9.6 Development of New Services

A proposal has been made to allocate frequencies adjacent to the 6/4 GHz bands to a new "low technology satellite service" (LTSS). This service would serve thin route or lightly trafficked systems using small earth station antennas. Although large satellite spacing of five degrees or more would be required to accommodate the use of low performance earth station antennas, it is anticipated that the LTSS could satisfy the needs of the developing countries by allowing them to implement relatively inexpensive earth stations.[36]

1.9.7 Intersatellite Links

The use of intersatellite links (ISLs) has also been cited as a way to reduce inhomogeneity between systems. Direct satellite-to-satellite relays, such as those mentioned in connection with the TDRS system, have been suggested as an alternative in facilitating coordination between systems when portions of the GSO are unavailable for satellites serving global or very large coverage areas. If the ground stations are more than about 8,000 kilometers apart, an intermediate ground station and a second satellite are necessary for communication. This uses twice as much of the GSO and radio spectrum capacity as a channel via a single satellite.

It has been proposed that satellites serving earth stations far apart in longitude — satellites with an unavoidably short service arc in the GSO — should be replaced by two satellites with direct interconnectivity, thus avoiding "double hops." Each of these satellites would have a long service arc, consequently a greater flexibility in the choice of orbital location. The use of ISLs could therefore facilitate coordination between global or large coverage area systems and domestic or small coverage area systems by allowing for reduced coverages. Such links could also use frequencies subject to high atmospheric attenuation — frequencies that are not suitable for satellite-to-ground communication and that are not subject to interference with terrestrial systems.[37]

Technically, the feasibility of using ISLs has been experimentally shown. In the short and medium term, however, use of ISLs would carry a large implementation and economic penalty when used to provide wide-area coverage because of the number of additional satellites and the technological sophistication required. Therefore, use of ISLs is not currently considered a viable option for the near future as a substitute for global beams.[38]

1.9.8 Multiple-Access Techniques

Various techniques are used to optimize use of the GSO and radio spectrum, including numerous multiple-access techniques. FSS networks have generally used the frequency modulation (FM) system for communications. This system is also used by terrestrial systems. Frequency division multiple access (FDMA) is used to allow a number of earth stations using different frequencies within the assigned band to gain access to a satellite transponder. Multiple-access techniques permit communication satellites to use the capacity of the satellite's transponders efficiently when numerous ground stations need simultaneous access to the satellite.

A time division multiple-access (TDMA) technique has been developed because the FDMA system becomes less efficient as the number of earth stations increases. In the TDMA system, digital signals and a sophisticated, signal-processing computer are used to allow the satellite to act as a sort of switchboard, rapidly switching transmissions between frequencies and between antennas. The satellite computer is able to process a great number of transmissions from multiple ground stations operating on various frequencies and transmitting bursts of signals on a sequential, non-overlapping basis. It does this by switching back and forth between earth stations in rapid sequence.[39]

Digital systems are able to use digital data compression techniques. In contrast to an FM-FDMA system, a TDMA system with digital speech interpolation (DSI) can transmit voice three times as efficiently. Using other digital data compression techniques, high fidelity television can be transmitted three times more efficiently also.[40] Other multiple-access techniques like code-division multiple-access and spread-spectrum multiple-access also are being developed and are proving highly beneficial for particular communication systems.[41]

1.9.9 Use of Higher Frequencies

A great deal of radio spectrum is available for use, especially above 10 GHz. This available spectrum can be used to satisfy demand for communication satellite frequencies. FSS systems increasingly use and will continue to use the higher frequency bands. Although the technology for using these higher frequencies is in early development, use of these bands may prove more attractive, both economically and technologically, than intensive, frequency reuse techniques.

Orbital congestion in selected areas also points to the use of higher frequency bands because use of higher frequencies permits greater antenna directivity, specificity, and discrimination. Better antennas allow closer satellite spacing and the accommodation of more satellites in the GSO. In addition, not as much terrestrial use is made of these high frequencies, thus reducing the possibility of terrestrial interference. Finally, the higher frequency bands tend to have wider bandwidths available. For instance, a 3,500 MHz bandwidth is available between 17 and 31 GHz.[42]

The current disadvantage to using some of these higher frequencies is that the present cost of implementing the systems is more than the cost of well-established systems. Thus, a greater demand still exists for the lower 6/4 GHz frequency bands of the FSS, especially from the developing countries.

1.9.10 Space Platforms and Multipurpose Satellites

Multipurpose satellites that combine two or more missions onto a single satellite have been proposed as a way to reduce the probability of collision in the GSO through a reduction in the total number of satellites. It has been suggested that a single satellite serving several countries could also cost each country less and could contribute less to congestion of the GSO than establishing satellites for each country, particularly countries with low demand. Multimission platforms, or "antenna farms," also have been suggested as a way to use the radio spectrum more effectively by permitting more flexible switching between ground networks, between geographic areas, and potentially through direct intersatellite or interplatform links.[43]

The disadvantage of using large space platforms for multimission purposes is that they present substantial problems of coordination and technical design. Another drawback to large, multipurpose platforms is a reliability problem until repair and maintenance in space becomes routine. Moreover, current and forseeable transportation systems are unable to launch these large platforms preassembled. Construction of such platforms appears to require the support of inhabited space stations. Thus, the advantage of using such systems is not yet clear. One alternative to space platforms that has been discussed is the employment of a cluster of satellites, designed to be colocated at the same orbital position without colliding. This would permit a gradual buildup of the constellation of satellites, through use of available launch vehicles, and economical replacement of any failed satellite.[44]

1.9.11 Regulatory Techniques

Absent a definitive technical answer, administrative and regulatory techniques have also been proposed to help ameliorate problems of orbital congestion associated with the use of the radio spectrum in the GSO. These techniques include international planning of the FSS, and new coordination procedures that incorporate concepts like "burden sharing" to facilitate and to implement the coordination of satellite networks.

It has been suggested, for instance, that the capacity to reposition a satellite be incorporated into its design. This would permit readjustment of satellite positions if necessary or appropriate to accommodate a new satellite system. Still, it would be difficult to readjust the positions of all satellites and their associated earth stations anytime a new satellite is placed into the GSO.

Currently, satellite networks are coordinated internationally on a bilateral, or country-to-country basis, when determining orbital positions for new satellites. This method helps to ensure that interference levels between systems are kept below permissible levels. Discontent has been expressed, however, particularly by the developing countries, over the use of the bilateral coordination process. As a result of growing concern over a possible orbit capacity shortage, the General World Administrative Radio Conference (GWARC) held in 1979 called for a special administrative radio conference to discuss and "guarantee in practice for all countries equitable access to the geostationary satellite orbit and the frequency bands allocated to space services."[45]

The first session of the two-session special conference was held in 1985 — the World Administrative Radio Conference on the Use of the Geostationary-Satellite Orbit and the Planning of Space Services Utilizing It (WARC-ORB-85). At this conference, proposals for planning and for new coordination techniques were put forth, with final determinations to be made on them at the second session of the conference to be held in 1988.

Long-term planning of the GSO for the fixed-satellite service in specific frequency bands was one proposal made at WARC-ORB-85. There was broad support at the 1985 conference by the developing countries for planning of space services. Such planning might guarantee to each country the use of a specific frequency or frequencies to be used by a satellite in a reserved orbital location.

Support for planning by the developing countries is prompted by an apprehension that the developed countries are pre-empting orbital positions in the GSO and the most desirable frequencies for the space services. Some developing countries believe that use of the current, bilateral satellite coordination process will not allow their anticipated future satellite communication needs to be met.

Another alternative proposed at WARC-ORB-85 was a new multilateral or multinational method for satellite coordination to augment the current bilateral process. It was decided by the conference that such an approach could be applied to the FSS bands that are not subject to an existing allotment plan. Unused "expansion bands" designated by the 1979 GWARC were recommended to the second session of the conference for incorporation into an orbital arc allotment plan for the FSS. This plan, to be established in 1988, may designate as much as 800 MHz of expansion bandwidth to each country for use in association with some portion of the usable orbital arc for their territory.

Periodic multilateral planning meetings were recommended by WARC-85 as one means for gaining access to the GSO and to radio frequencies not subject to the arc allotment plan. These international meetings would adjust the plan as necessary and would mediate any necessary "burden sharing" between administrations. Burden sharing was proposed as a way to help implement the various technical and operational problems associated with harmonizing the satellite systems in the GSO. This concept recognizes that the various steps toward harmonization can be a "burden" to be shared between existing and new networks.[46] The concept of burden sharing includes various elements like satellite relocation, elements that would be shared both operationally and economically between administrations concerned. The possible implementation of burden sharing and of multilateral planning meetings are also to be considered in 1988. (See Table 1.4.)

1.10 CONCLUSION

These administrative proposals have been made with the intent of contributing to the growth of the available capacity of the GSO and radio spectrum. Such proposals represent the kind of international cooperation that has occurred over the years to accommodate or conform to the limits of the natural and technological resources that are associated with radio and space communication.

The following chapters demonstrate how it has proven feasible through limited, flexible, and judicious planning, combined with rapid technological improvement and continued international coordination and cooperation, to ameliorate most major problems of congestion and radio interference that might otherwise arise in uncoordinated circumstances. The mature state of the international law and regulation of international telecommunication has provided the framework for modern space communication law and regulation.

As we have seen, the characteristics of the radio spectrum make space communication innately like radio communication. This helps to explain why the origins and lineage of space communication law and regulation are over one hundred years old, and date back to the beginnings of electronic communication.

Table 1.4 Potential Utilization Improvement Factors for Communication Satellites in the GSO
Source: "Efficient Use of the Geostationary Orbit," UN Doc. A/CONF. 101/BP/7, 1981.

Purpose	*Technology*	*Change*	*Approximate Improvement Factor*	*Notes*
Reducing Intersatellite Spacing	Antenna Beamwidth Reduction	Larger Antennas 30m vs. 11m	2	Increased Earth Station Cost
	Antenna Sidelobe Reduction	7.5 dB Reduction from CCIR Standard	2	New Antenna Technology
	Station-Keeping	$\pm 1°$ to $\pm 0.1°$	1.8	Recent Satellites Have $\pm$ 0.1° Capability
	Optimum Positioning	Multilateral Coordination vs. Bilateral Coordination	1.5	Complex Coordination and Increased Satellite Cost
	Modified Orbits	Inclined and Eccentric Orbits	2–4	New Station-Keeping Technology
Increasing Effective Bandwidth	Available Frequencies	Bandwidth Increase from 1000 MHz to 5500 MHz	5.5	New Communications Technology
	Beam Separation	Global Coverage to 6 Spot Beams	6	Limited by Concentration of Traffic
	Polarization	Unpolarized to Dual Polarization	2	Implemented on Some Recent Satellites
	Uplink-Downlink Assignments	Alteration of Uplink-Downlink Frequencies	1.8	Interference and Coordination Problems
	Intersatellite Links	Elimination of Double-Hops	1.1	New Satellite Technology
Increasing Communication Efficiency	Modulation and Multiple Access	FM/FDMA to DSI/TDMA	3	Increased System Cost
	Data Compression	Voicc and TV Signal Compression	3–6	New Communications Technology

REFERENCES

1. Correspondence to authors from Stephen E. Doyle, Esq., chairman FCC Space WARC Advisory Committee for the ITU World Administrative Radio Conference on the Use of the Geostationary-Satellite Orbit and the Planning of the Space Services Utilizing It, June 16, 1987.
2. *ITU Radio Regulations,* 1982, Art. 1.
3. President's Task Force on Communications Policy, *Final Report,* Washington, DC: US Government Printing Office, 1968, Chap. 8, p. 2, quoted in Levin, Harvey J., *The Invisible Resource Use and Regulation of the Radio Spectrum,* Baltimore: Johns Hopkins Press, 1971, p. 15.
4. Hewitt, Paul G., *Conceptual Physics A New Introduction to Your Environment,* Boston: Little, Brown and Company, 1977, p. 275.
5. 1 Kilohertz (kHz) = 1,000 Hz; 1 Megahertz (MHz) = 1,000,000 Hz; 1 Gigahertz (GHz) = 1,000 MHz.
6. Codding, George A. Jr., and Anthony M. Rutkowski, *The International Telecommunication Union in a Changing World,* Dedham, MA: Artech House, 1982, p. 51.
7. *Final Acts of the World Administrative Radio Conference, Geneva, 1979,* Vol. 1, Art. 8.
8. Levin, *supra* note 3, at 20.
9. *Final Acts of the World Administrative Radio Conference for Space Telecommunications, Geneva, 1971,* Annex 3, Art. 5.
10. *International Telecommunication Convention, Nairobi, 1982,* Annex 2, Para. 2003.
11. *Ibid.*, Art. 4.
12. See Levin, *supra* note 3 for a detailed description of the properties of radio and the effects upon it.
13. "Efficient Use of the Geostationary Orbit," *United Nations Document A/CONF.101/BP/7,* January 16, 1981, p. 5.
14. Interview by telephone with Russel D. Conquist, spacecraft systems engineer, Westinghouose Electric, October 1986.
15. "Efficient Use of the Geostationary Orbit," *supra* note 13, at 18.
16. *Second Report of the Advisory Committee for the ITU World Administrative Radio Conference on the Use of the Geostationary-Satellite Orbit and the Planning of the Space Services Utilizing It,* 1985, pp. 2,4, (herein cited as *Second Advisory Report*).
17. World Administrative Radio Conference on the Use of the Geostationary-Satellite Orbit and the Planning of Space Services Utilizing It, First Session, Geneva, 1985, *Report to the Second Session of the Conference,* Chap. 2.1, p. 7, (herein cited as *Report to the Second Session*).

18. "Efficient Use of the Geostationary Orbit," *supra* note 13, at 5.
19. *Ibid.*, p. 12.
20. *Ibid.*, p. 13.
21. *Ibid.*, p. 14.
22. *Ibid.*, p. 6.
23. *Ibid.*
24. *Ibid.*, p. 18.
25. *Ibid.*, p. 15.
26. *Report and Order in the Matter of Licensing of Space Stations in the Domestic Fixed-Satellite Service,* (Two Degree Spacing Decision), 93 FCC 2d 1261, 1983.
27. "Efficient Use of the Geostationary Orbit," *supra* note 13, at 7.
28. *Ibid.*, p. 20.
29. "Current and Future State of Space Technology," *United Nations Document A/CONF.101/BP/2,* March 16, 1981, p. 12.
30. "Efficient Use of the Geostationary Orbit," *supra* note 13, at 20.
31. *Ibid.*
32. *Ibid.*, p. 8.
33. *Report to the Second Session, supra* note 17, Chaps. 3.4.2.4.2.; 3.4.2.4.4.
34. *Ibid.*
35. *Ibid.*, Chap. 3.4.3.4.4.3.
36. *Second Advisory Report, supra* note 16, at 10.
37. *Report to the Second Session, supra* note 17, Chap. 3.4.2.5.3.
38. *Ibid.*, Chap. 3.4.2.5.3.3.
39. "Efficient Use of the Geostationary Orbit," *supra* note 13, at 22.
40. *Ibid.*
41. *First Report of the Advisory Committee for the ITU's World Administrative Radio Conference on the Use of the Geostationary-Satellite Orbit and the Planning of the Space Services Utilizing It,* 1983, pp. 3-18 to 3-19.
42. *Report to the Second Session of the Conference, supra* note 17, Chap. 2.4.4.8.
43. "Efficient Use of the Geostationary Orbit," *supra* note 17, at 23.
44. "Current and Future State of Space Technology," *supra* note 29, at 17.
45. *Final Acts of the World Administrative Radio Conference, Geneva, 1979,* Vol. 2, Resolution 3.
46. *Report to the Second Session of the Conference, supra* note 17, Chap. 3.4.3.5.3.1.

Chapter 2
The International Telecommunication Union and the Regulation of Radio: Origins and Evolution

2.1 INTRODUCTION

Growth in the use of satellite communication is intimately involved with more effective use of the geostationary satellite orbit (GSO) and the radio spectrum. Yet, radiocommunication was a comparatively late development in international communication. It was the use of the electric telegraph that spawned the necessity for an international organization that could facilitate cooperation among the European nations that used telegraphy for international communication.

This chapter demonstrates that the bedrock of institutions and principles upon which the law and regulation of international space communication are based was formed in response to several new technologies over a long period of time. The birth of these institutions and principles can be traced to the international response to the need for establishment of regulations that would help the development and coordination of early international telegraph networks.

Half a century after the development of the electric telegraph, the new telephone also was regulated by the successor of the international institution that emerged as the principal forum for telegraph regulation. Ultimately, these early organizations led to the formation of the international body responsible for the regulation of space communication — the International Telecommunication Union (ITU).

After the electrical telegraph was proven practical, telegraph lines quickly became established in the major, more developed nations of the world. A line between London and Birmingham was operating in England in 1837.[1] Samuel F.B. Morse, financed by a grant from the United States government, ran the first US telegraph line between Washington, D.C., and Baltimore, Maryland. On 24 May 1844, he tested his invention. "What hath God wrought?" flashed between the two cities.[2] Public ownership and operation of the line, however, was not self-supporting. Consequently, the first US telegraph line, which was originally placed under the US Postmaster General, was sold in 1847 to private interests.

France, Belgium, Austria-Hungary, Italy, Switzerland, and Russia all had telegraph lines in operation by the early 1850s. These lines not only provided for rapid communication but were also indispensable assets to the safe and rapid operation of another invention introduced during this period — the railroads. The world had entered an era of industrialization, rapid transportation, and instantaneous electronic communication.[3]

Not long after the establishment of the telegraph lines in the major nations of the world, the first international telegraph agreement was made. In 1849 an agreement was struck between Prussia and Austria in an attempt to coordinate and to establish some sort of uniformity in the operations of telegraph systems, such as standards, codes, and tariffs. This loose-knit association, known as the Austro-German Telegraph Union, was later joined by other countries of Central and Eastern Europe.[4]

In 1851 France and Belgium also established a cooperative venture to aid in the regulation and coordination of electrical communication. This union of States, known as the West European Telegraph Union, was later joined by other nations of Western Europe.[5] These early international organizations, driven by technological change, were formed to coordinate and to regulate international telecommunication and were the prelude to the international union that emerged to become the direct predecessor of the present International Telecommunication Union.

2.2 THE INTERNATIONAL TELEGRAPH UNION

2.2.1 The 1865 Paris Telegraph Conference

In 1865 the Austro-German and the West European Telegraph Unions met in Paris and merged, forming the International Telegraph Union. Not only did the Paris Telegraph Conference create a new international body to oversee the burgeoning development of the new electrical communication networks, but it also drafted a treaty that established uniform telegraph standards and adopted the Morse code for use in international lines.[6]

Two very significant documents emerged from the Paris Conference. These were the International Telegraph Convention and annexed Telegraph Regulations. These two documents are historically important because they established an intra-European telegraph network and the International Telegraph Union to supervise the network.

As part of the organizational design of the new International Telegraph Union, periodic conferences of official delegates were provided for in the new International Telegraph Convention to serve as the formal mechanism by which the Convention could be revised when necessary. It was the task of conference delegates to revise the regulations in order to keep the international telegraph network up to date.[7]

2.2.2 The 1868 Vienna Telegraph Conference

Vienna was the place where the second Union Conference was convened in 1868. This conference completed the original structure of the International Telegraph Union by creating the International Bureau of Telegraph Administrations.[8] The new body, known as the Berne Bureau, because it was located in the Swiss capital of Berne, became an indispensable arm of the Union and performed all of its routine administrative work. In essence, the Berne Bureau functioned as a modern-day secretariat.

With the exception of a name change in 1906 to the International Bureau of the Telegraph Union, the Berne Bureau functioned under successive conventions for eighty years. In 1948, under a revised ITU structure, tasks formerly performed by the International Bureau were apportioned between the new International Frequency Registration Board and the General Secretariat.[9]

The Berne Bureau was staffed, administered, and directed by the Swiss, with a class-unit system determining how much each member of the Union paid for Union financial support — a system similar to today's system that essentially allows each administration to select its own contributory class, thereby determining what it will pay. The Bureau was charged with collecting all information related to international telegraph and with collecting general statistics. The Bureau published current telegraph rates and a journal of telegraph matters and was charged with conducting special studies when so directed.[10]

2.2.3 The 1871 Rome Telegraph Conference

The third conference of the International Telegraph Union was held in Rome in 1871. Some of the items on the agenda of this conference included submarine telegraphy and various organizational matters, which resulted in important changes in the organization of the Union. For the first time an American, Cyrus Field, attended a Union conference. Field, however, the entrepreneur responsible for laying the first trans-Atlantic cable, represented a non-US, private company. It is interesting to note that Field asked the conference to consider the universal protection of submarine and land cables of all nations during war.[11]

The Plenary Assembly of the conference decided, however, that such a resolution was outside the scope of its authority. The point was made at the meeting that the improvement of telegraph matters, not international law *per se*, constituted the only jurisdictional concerns of the conference.[12]

The Rome Conference granted private companies the right of representation at all meetings. They were not, however, given the right to vote. Ten different private companies attended the conference. The issue of private business representation had been raised because many telegraph lines were operated by private companies. Because the companies were under the jurisdiction of their sovereigns — members of the International Telegraph Union who were signatories to its Convention and thus bound by its Convention and Regulations — it was felt that the private companies should have the right to offer advice on the regulations to which they were subject.[13] To this day, Recognized Private Operating Agencies (RPOAs) and Scientific or Industrial Organizations (SIOs) contribute and play a substantial role in the formation of international telecommunication regulation.

2.2.4 The 1875 St. Petersburg Telegraph Conference

The Rome Conference also charged the Berne Bureau with drawing up a draft convention for the St. Petersburg Telegraph Conference scheduled for 1875. The convention that finally emanated from the 1875 conference was based on the preliminary draft that had been called for by the Rome Conference and did not much differ from it. A distinct feature of this convention placed all elements of a permanent nature in the convention and those subject to periodic revision in a set of annexed technical regulations.[14]

The St. Petersburg Convention remained essentially unchanged until 1932, almost sixty years after it was drafted.[15] It is interesting to note here that several articles from the 1875 St. Petersburg Telegraph Convention were eventually incorporated into the 1906 Berlin Radiotelegraph Convention and its annexed Radio Regulations. These documents exist as the direct predecessor of today's international regulations for space radiocommunication.[16]

2.2.5 The 1885 Berlin Telegraph Conference

With the rapid development of telegraph technology, the new telephone invention was looked upon as an inefficient competitor. In fact, the matter of telephone regulation was not a popular subject. Nevertheless, the matter was discussed at the 1885 Berlin Telegraph Conference. As a result of opposition by various European governments who felt that strict regulation would deter the development of the new telephone technology, only a short passage related to telephone was added to the telegraph regulations. Standards were set only for time-unit charges and the length of conversations allowed when other requests were being made for the line. Individual governments retained almost complete freedom in matters related to telephone regulation.[17]

It was not until 1903 that telephone regulation became an important part of the work of the Union. At the London Telegraph Conference of that year, a special committee was designated for the sole purpose of drafting regulations for the international telephone service. Telephone use by that time had grown to the point where definite rules were necessary for effective operation. Consequently, the regulations devoted to telephone were expanded from a mere five paragraphs in the 1885 Regulations to 15 articles consisting of more than 60 paragraphs in the 1903 Regulations. Thereafter, the International Telegraph Union could have been referred to as the International Telegraph and Telephone Union.[18] Telephone service today constitutes a significant portion of international satellite communication traffic.

2.3 THE 1903 BERLIN PRELIMINARY RADIO CONFERENCE

In 1903 the scope of international telecommunication regulation was further expanded by the results of the Berlin Preliminary Radio Conference (Preliminary Conference Concerning Wireless Telegraphy), which, though not an official activity of the International Telegraph Union, laid the basis for a binding treaty on radiocommunication that was completed in 1906.[19]

At the invitation of the German government, nine countries, including the United States, met in Berlin in 1903 to discuss matters related to the international regulation of radio.[20] It became obvious that regulations were needed for humanitarian reasons and to enhance safety of life at sea.

In those early days of radio development, the Marconi Wireless Company held agreements with the Italian government and with the British Lloyd Company that established its exclusive right to install radio equipment on ships and to provide company technicians for its operation. The company was also given the right to refuse to communicate with any other radio station that did not use the Marconi Company equipment. In essence, the Marconi Company held a virtual monopoly on radiocommunication between ships and between ship and shore.

As a consequence, incidents related to safety of life at sea were frequent. One such incident occurred when a US ship that was ordered to be on the lookout for a derelict ship endangering navigation in the Atlantic shipping lanes radioed a German ship fitted with Marconi equipment. The German ship, forbidden by contract to communicate with any ship using other than Marconi equipment, replied, "Can not [sic] give it to you. Not allowed to work with you." Infringement of these contractual rules laid down by the Marconi Company could result in instant dismissal of the employee concerned.[21]

Other less dangerous incidents occurred also, such as when the Prussian Prince Henry was within 100 miles of the American coast and tried to send a message to the US President Theodore Roosevelt announcing his pending arrival. The Prussian Prince was unable to communicate with shore because an outward-bound ship was conversing with shore and refused to relinquish the frequency.[22] Several naval incidents of this nature prompted the calling of the first international conference on radio, to which many countries concerned with the unobstructed future use of radio responded.

The Final Protocol produced by this 1903 Preliminary Radio Conference contained important regulatory provisions that were to be taken back to the respective governments for review. The Protocol extended certain provisions of the 1875 St. Petersburg Convention to radio. The 1875 Convention contained various provisions, one of which was related to the universal right of everyone to correspond by telegraphy. The Convention also dealt with secrecy of communication, the prerogatives of the State, and the cost of communications, which was determined by the number of words transmitted.[23]

The Final Protocol of the 1903 Berlin Preliminary Conference also provided that priority of communication be given to ships at sea and that radio stations should operate without causing interference to other stations.[24] Although there was agreement at the conference that distress signals should be given priority, the decision as to which distress signal to use was not resolved. The British suggested the use of CQD as a distress call as they used this signal on their railroads. The Germans wanted SOE. Consequently, consensus could not be reached.[25]

Great Britain and Italy made declarations reserving their respective governments' rights not to comply with various provisions of the Final Protocol. The principles put forth in this document became the basis for the regulation of radio, however, and stood as the draft for the 1906 Berlin Radiotelegraph Convention, the first document of international radio law.[26]

2.4 THE INTERNATIONAL RADIOTELEGRAPH UNION

At the invitation of the German government, all interested maritime governments were invited to attend a conference in Berlin in 1906 to draw up a Convention and Radio Regulations based upon the Final Protocol adopted in 1903. During the interim period of 1903 to 1906, however, increasing friction between the United States and the British Marconi Company over maritime radio operations persisted.[27]

During this period, the Marconi Company continually ignored the recommendations of the 1903 Protocol by refusing the messages transmitted by rival systems. For example, the English company had radio equipment in the US lightship *Nantucket* off the US coast. Marconi radio operators on the ship consistently refused communications from ships equipped with other systems that came into the harbor. As a result of complaints by the Germans, the United States told the Marconi Company to accept communications from ships installed with equipment from other companies. The Marconi Company refused and was subsequently asked to vacate the lightship. Because of these and other naval incidents, the United States aligned herself with Germany at the 1906 Radiotelegraph Conference.[28]

2.4.1 The 1906 Berlin Radiotelegraph Conference

The 1906 Berlin Radiotelegraph Conference was composed of delegates from the most important maritime powers of the world. Delegations from 29 states were represented at this international conference that drafted the Berlin Radiotelegraph Convention and an annexed set of Radio Regulations.[29]

From the provisions of the 1906 Convention emerged an organization that became known as the International Radiotelegraph Union. The Berne Bureau was designated to serve as its central administrative organ.[30]

Although the convention did not formally establish an organized Radiotelegraph Union, the various administrations that were involved in and worked together on matters concerning radiocommunication developed into a *de facto* organization that was commonly referred to as the International Radiotelegraph Union.[31]

Although the primary objective of the Berlin Radiotelegraph Conference was to facilitate maritime communications by requiring intercommunication between stations no matter what equipment was used, the conference was no more able to change the monopolistic practices of the Marconi Company than was the 1903 Conference. Still, the conference did succeed in establishing many of the fundamentals of radio regulation that exist today.[32]

Two important regulatory principles were incorporated into the 1906 Convention. These principles were the regulatory concepts of classification of radio services and the obligatory notification of national radio station frequency assignments. The classification of radio services entails specifying frequency bands that are allocated to specific radio uses.[33]

This regulatory principle of allocating frequency ranges to specific services, and that of notification of national radio station assignments to a central recording agency, set the precedent in 1906 for the case-by-case, *a posteriori*, first-come, first-served process for establishing international recognition of specific uses of radio that is still in place. This method usually relies on a notification-registration process and necessitates coordination among concerned parties. As noted in the Introduction to this book, however, the question of the equitable validity of this method of attaining rights to radio spectrum use is highly contested and is currently being challenged in international radio conferences, with an *a priori* method of attaining recognized uses being sought. This latter method usually relies on a negotiated plan that is based on general criteria for achieving equity among all the parties.

To carry out the requirement for the notification of radio station frequency assignments, the Berne Bureau was designated as the arm of the Union that was to act as the repository for and the publisher of the notifications. The Bureau was to compile a list of radio stations from these notifications. The list was first published as The Frequency List of Radioelectricity Stations (The Berne List), and later became the International Frequency List.[34]

This notification and recordation process encompassed many basic elements that still exist today, and had the effect of establishing priority rights to use specific radio channels in specific areas. The Bureau performed these functions until the reorganization of the ITU in 1947 placed these responsibilities under the jurisdiction of the newly created International Frequency Registration Board (IFRB).[35]

The 1906 Conference also established an important organizational concept. Henceforth, meetings at regular intervals were to be held to revise the Convention and the Radio Regulations.[36] Moreover, in addition to the regulatory and organizational concepts embodied in the 1906 Radiotelegraph Convention, the international SOS distress signal also was formally adopted.[37]

2.4.2 The 1912 London Radiotelegraph Conference

As has been noted, certain articles from the 1875 St. Petersburg Convention were incorporated into the 1906 Radiotelegraph Convention and made applicable to international radiocommunication. This procedure was followed also in the drafting of the Regulations of the 1912 London Radiotelegraph Convention. Provisions concerning radiotelegraph rates, however, also were included in the 1912 Convention.[38]

Because rate provisions were incorporated into the 1912 Radiotelegraph Convention, both the United States and Canada made formal reservations to the Convention concerning such provisions. Both administrations also refused to belong officially to the International Telegraph Union because rate provisions were dealt with in its convention.[39]

On the grounds that its telegraph and telephone networks had never been nationalized, the United States refused to join an organization that regulated their use. It pointed out that since most of its telegraph, telephone and cable systems belonged to private companies, the US government needed their approval before it could undertake any obligations. Therefore, the United States never became a member of the International Telegraph Union.[40]

Rate regulation historically had been a thorny subject to the United States, and the inclusion of items related to it was an important deterrent to the United States in signing the 1912 Radiotelegraph Convention. Private companies owned and operated the telegraph lines in the United States and the transmission of radiotelegrams also was done by commercial interests.[41]

The United States maintained that the matter of rates was primarily a management problem, which could be settled by ordinary business and economic principles and usages. The US position was that matters regarding rates required frequent change and could not be adequately dealt with by general constitutional provisions that are rigid and slow to change. Rate matters were considered to be the province of private management, which, it was felt, could efficiently handle rate matters in a quick and flexible manner. In the United States, managerial matters are not covered by positive constitutional law because a constitution is deemed appropriate only for establishing the legal context, and amendments to a constitution require too much time and delay.[42]

It took nearly 85 years before the United States finally signed a convention's annexed Telegraph Regulations in 1949 for the first time. This occurred only after substantial changes to the Regulations had been enacted. Moreover, it was not until 1973 that the United States signed the Telephone Regulations for the first time.[43]

The primary reason the United States decided to sign the Telephone Regulations in 1973 was that the Administrative Telegraph and Telephone Conference of that year transferred most of the telegraph and telephone provisions from treaty-status Regulations to Recommendations that are the jurisdiction of the present ITU consultative committee that handles wire communication — the International Consultative Committee for Telegraph and Telephone (CCITT). As a consequence, the time-consuming formalities of treaty ratification became unnecessary and the provisions could now be constantly reviewed and amended as necessary by study groups without having to wait for an administrative conference to amend the treaty regulations.[44]

Historically, as far as rates were concerned, the United States was successful as early as 1927 in proposing that the Radiotelegraph Regulations be divided into two parts: "one containing general principles that the United States could adhere to as a Government, and the other containing all the provisions, such as rates, accountability, *et cetera*, that the United States usually left to the competence of the private companies."[45]

All articles dealing with radio charges were placed in a supplemental section called the Additional Radio Regulations. In this way objecting nations like the United States and Canada were able to accept without reservation the general Radio Regulations while refusing to ratify the Additional Regulations. From 1927 until they were eliminated in 1979, the United States was never bound to the Additional Regulations.[46]

Of course, matters other than rates were also addressed at the 1912 Radiotelegraph Conference. This conference was significant in that, for the first time, the Marconi people agreed to end their practice of refusing to communicate with others not equipped with Marconi equipment. This agreement came against the backdrop of the Titanic disaster, which occurred two months before the conference opened, and gave impetus and urgency to the work of drawing up a new set of regulations for the better utilization of radio on ships.[47]

Various proposals were made at the conference to enhance safety of life at sea. The most important of these proposals included provisions dealing with the authority of the ship's captain over the company who operated the radio service on board, and with the necessity for auxiliary equipment for emergency use in case the main power of the ship failed. The need for a constant watch, especially on board the larger passenger ships, was also enunciated. Due to opposition based primarily on supposed economic consequences should such proposals become mandatory obligations, however, the decision was made not to establish any absolute rules at that time.[48] Intercommunication between stations was made obligatory, however, and the priority of distress calls was provided for in the 1912 Convention.[49]

It was also at the 1912 Radiotelegraph Conference that, for the first time, economic aspects of frequency management were linked to the principle of efficiency. The desire for more efficient and economical use of the spectrum was driven by improvements in technology and the desire to avoid harmful interference. This was recognized when the concept of the pure or sharp wave was considered. At the time, ships were using spark transmitters that emitted a type-B radio wave. Although these transmitters were cheap, their emissions not only occupied a broad bandwidth, but they tended to wander off any assigned frequency, consequently causing interference to other services or stations. The 1912 Convention required that the pure wave be used as much as possible. Spark transmitters, "in which the waves sent out are obtained by means of sparks directly in the aerial," were unauthorized except in cases of distress.[50]

The next international conference to determine radio regulation was set for 1917. World War I intervened, however, and it was cancelled. Consequently, the development of international radio regulation after the war is especially interesting in that deliberations reflected the changing sentiment and the changing technology.

2.5 POST-WORLD WAR I NON-UNION MEETINGS

During World War I all governments invested heavily in radio research and built numerous radio stations. Radio became an important aspect in naval and in ground operations. Very notable achievements in aircraft radiocommunication also were made by the armies and the navies of combatant governments. Because of developments related to cable that occurred during the war, the states of the world turned to radio for their communication needs. As a result, the number of coast radio stations nearly doubled and the number of ship stations more than quadrupled by 1920.[51]

The cable problems that prompted this surge in radio use are of historic interest. At the outbreak of the war, Germany was one of the major cable owners of the world and an important news center. The English cut the German cable to New York during the war and diverted it in two places. The English Channel end went to England and the other cut, made 600 miles out at sea, was diverted to Halifax, Canada. Thus, the cable that once went from Germany to the United States now became an English-Canadian cable. The French cut another German-United States cable in the English Channel and diverted the German end to Brest, France. The American end was moved from the German landing in New York to the French landing site there. Consequently, Germany lost her cables during the war and the United States, even as a victor, practically lost the use of her cables.[52]

2.5.1 The Radio Protocol of 1919

At the Versailles Peace Conference, held from 18 January to 28 June 1919, Germany relinquished her claims to the cables, turning them over to the Allied and Associated Powers. A five-power international committee was appointed by the Supreme Council of the Allies to consider applicable international law to the disposition of the German cables. A series of secret meetings attended by the Allied Powers — the United States, Great Britian, France, Japan, and Italy — occurred soon after the committee was established. Although the meetings were called to consider certain aspects of the peace negotiations, it seems that neither these meetings nor the international committee were able to resolve the problem of the submarine cables. The meetings, however, did produce a resolution calling for an international conference for the drafting of an international convention for all electrical communications.[53]

This resolution, signed in conference on 25 August 1919 by the Inter-Allied Radiotelegraphic Commission, became known as the EU-F-GB-I Radio Protocol of 1919. The letters stood for the initials in French of the countries that signed the protocol — the United States, France, Great Britain, and Italy. The Commission had been directed by the Supreme Council of the Allies to prepare a draft set of provisional amendments to the 1912 Convention for the next radiotelegraph conference.[54]

The Commission produced many of the currently existing approaches to frequency management and expanded the rudimentary allocation scheme that was set forth in the 1912 Regulations into what became the forerunner of the current ITU Table of Frequency Allocations. The radio spectrum from zero to 12,000 meters was allocated to radiotelegraphic services. (The conversion to frequency from metric wavelength to measurement to signify the boundaries of service bands was not made until 1921.) There were four categories of service. Those categories were: mobile, fixed, military and special. Further restrictions were also proposed on the use of old spark equipment.[55]

Another proposal submitted to a meeting of the Allied radio experts concerned the formation of an International Technical Committee that would serve to advise on matters related to the radiotelegraphic services.[56] This proposal for an advisory committee, however, did not come to fruition until 1927 when the international consultative committees were created.

2.5.2 The 1920 Washington Preliminary Conference

It should be noted that initially World War I produced a world climate receptive to the creation of international arrangements and organizations like the League of Nations. Therefore, at the conference called for by the 1919 Protocol — the 1920 Washington Preliminary World Conference on Electrical Communications — it was not surprising that the Allied Powers considered dramatic revision of the 1912 London Radiotelegraph Convention. Within the context of revising this Convention, it was proposed that both the Telegraph and the Radiotelegraph Conventions be integrated so as to create a Universal Electrical Communications Union.[57]

It should also be noted, however, that the work of many post-war telecommunication bodies was not adopted by the entire international community. Nevertheless, the Berne Bureau continued to operate and eventually gathered together and edited the work of the Inter-Allied Commission, the 1920 Preliminary Conference, various domestic commissions that studied the work of the Preliminary Conference in 1920-21, and of the Technical Committee meeting in Paris in 1921. The work product from these meetings subsequently served as the basis for proposals to the 1927 International Radiotelegraph Conference in Washington.[58]

The 1919 Radio Protocol urged interested nations to consider further revision of the 1912 Convention at the 1920 Washington Preliminary Conference. Preconference recommendations also were drafted by private companies in the United States and directed to conference delegates. Included in these were proposals from representatives of the press, who emphasized the necessity for a free flow of news. It was also suggested that special rates be granted to the press. The US delegation took these proposals to the 1920 conference.[59]

Since similar recommendations for a free flow of information and special rates for news are also made today in connection with space communication, it is interesting to note the different world conditions that existed to spawn such proposals. In the aftermath of World War I, England controlled nearly all existing cable systems. Numerous complaints were registered at the time by the US press. Complaints concerned two disadvantages of the system in use. High rates were charged and cable messages were routed to the Orient either through Japanese or British agencies. The US press claimed that its messages going through these agencies were censored and resulted in distorted, sensational messages at the terminal point. It was also charged that messages going through England were still being inspected by the British secret service even after the war had ended.[60]

As the conference opened in 1920, at the invitation of the United States, it had three main objectives to accomplish: to dispose of the German cables equitably; to establish a principle of equal opportunity in the use of the cables; and to permit the uncensored distribution of news at fair prices throughout the world.[61]

The cable problem was so important at the 1920 meeting that radio was only a secondary issue. Of the five working subcommittees established at the conference, however, at least two were primarily concerned with radio. These were the bodies charged with studying the technical aspects of the 1919 Radio Protocol and the formation of a Universal Electrical Communications Union. The other working subcommittees of the conference looked at issues related to the disposition of the German cables, international cable and radio law, cable landing rights, and the improvement of communication facilities between nations that attended the conference — the Allied and Associated Powers.[62]

The issue of the disposition of the German cables was never fully resolved at the 1920 Conference. After many hours of deliberations, however, a temporary agreement was made. After 1 January 1921 the net income from the ex-German cables was to be held in an escrow account. This arrangement would be reviewed if a permanent agreement had not been entered into by 15 March of the same year. A partial settlement did result, however, where England retained the German-New York line she had diverted during the war. Italy, France, and Japan all received other pieces of the cable, but the cable position of the United States remained worse than before the war.[63]

If these accomplishments were discouraging, the results of the work related to radio were much more optimistic. A proposal establishing a Universal Electrical Communications Union that would equitably allot radio frequencies among all nations of the world was adopted by the 1920 Conference. The creation of this Union, which would handle all telecommunication matters, would result from the amalgamation of the Conventions of the International Telegraph Union and of the *de facto* Radiotelegraph Union. A draft Convention for a Universal Union was drawn up, specifying that "coefficients of distribution" would be used as criteria at Union conferences in determining the frequency allocations to specific countries. These coefficients comprised various factors: population, extent and importance of colonies, extent of commerce, and the need for international communication. Prior rights were not to be granted in perpetuity, but for ten years or as dictated by Union conferences. Thus an *a priori* concept of planning that promoted a rights vesting approach had its genesis in 1920. The draft Convention also stipulated that once a prior right to a frequency band was granted a state, that state would have to make good use of it or suffer limitations imposed upon that use.[64]

The elements that form the basic structure of the present International Telecommunication Union were forged at the conference. Matters considered at the 1920 Washington Preliminary Conference included:

- the allocation of frequency bands among services;[65]
- the establishment of a distinct consultative organ within the Universal Union to advise on radio technical matters;
- the establishment of a governing council that would serve in the interim between conferences of the Union;
- the establishment of a Central Bureau to conduct administrative activities; and
- the recognition that private enterprises could deal directly with the Central Bureau.[66]

2.5.3 The 1921 Paris Technical Committee

A Technical Committee on International Radio-Communication also was established by the 1920 Conference and met in Paris during the summer of 1921. A major objective of this conference was to revise the technical parts of the Washington draft convention. Several unresolved questions had been referred to the Technical Committee by the 1920 Conference, including the issue of equitable allotment and the kind of administrative scheme that should be used for vesting rights in frequency usage. No consensus could be reached, however, and these questions were left unresolved.[67]

Probably the most important work of the 1921 Technical Committee was the drafting of a table of frequency allocations between 60 and 150 kHz. Frequencies were allocated to the mobile, military, fixed, and special services that included radio beacons, direction finding stations, meteorological, and other services.[68] As previously noted, however, the work of this and of other post-war meetings was never put into effect, but did serve as a basis for later conferences.

Indeed, an important result of the 1921 Technical Committee was to set in motion the preparation necessary to hold the next international radiotelegraph conference.[69] The period after 1921, however, was a time when the strong sense of internationalism and cooperation that existed right after the war began to decline. Various concepts like those for a League of Nations and for *a priori* planning of international use of the radio spectrum — concepts that had enjoyed the support of the administration of US President Wilson — lost support during the 1920s in the wake of the decline of internationalism.

2.6 THE 1927 WASHINGTON RADIOTELEGRAPH CONFERENCE

In 1927, at the invitation of the United States, the first post-war conference of the Radiotelegraph Union was held and met in Washington. Delegates from over 80 countries, not including the USSR, participated in the Washington International Radiotelegraph Conference of 1927. Sixty-four private companies and private and international organizations were also represented at the conference.[70]

The USSR was not invited to the conference by the US government as the latter did not politically recognize the Soviet State. The USSR sent a letter of protest to the Berne Bureau, which the conference chairman, US Secretary of Commerce Herbert Hoover, read before the Plenary Assembly. Apparently, there was no overt reaction to the letter by the Plenary Assembly, which listened in silence. As a result of not having been invited to the 1927 Conference, the USSR did not consider herself bound by any decisions reached in Washington.[71]

Many substantial revisions to the 1912 Convention were adopted by the 1927 Conference. A new Radiotelegraph Convention and new General Regulations were drawn up. In addition, as has previously been noted, the US proposal for an Additional Regulations, or Supplementary Regulations, were added. These Additional Regulations concerned matters like rates. The addition of these separate regulations satisfied the needs of the United States to find a way by which it could adhere to the Radiotelegraph Convention and the major portion of the General Regulations without being bound to provisions the United States felt were only within the province of its private companies to decide.[72]

The 1927 Conference also established the use of English as a second, unofficial language to be used for discussions. The official documents were still to be distributed in the official language of French, but thereafter, English versions of service documents were also distributed.[73]

The scope of the Radiotelegraph Convention was greatly increased at the Washington Conference. Prior to 1927, radio conventions dealt primarily with public maritime correspondence. The new Convention was drafted to include all public radiocommunication stations and many special services not used for public correspondence. These included services such as the meteorological and the radiobeacon services. The 1927 Convention also enlarged upon a clause in the 1912 Convention related to the secrecy of correspondence. In 1912 this clause referred only to the maritime-mobile service. The Washington Conference expanded the provision to include the operator of any transmitting station. A general obligation was imposed upon all contracting administrations to submit to their respective legislatures measures that would more clearly enunciate the principle of protection for security of correspondence.[74]

The principles of efficiency and economy, raised by the continuing issue of interference caused by the use of spark sets that emitted type-B radio waves, was again addressed at the Washington Conference. A 1 January 1940 date was set after which time the use of this type of radio wave on all frequencies would be prohibited. An exception was made, however, for low-power transmissions below 300 watts.[75]

A major achievement of the Washington Conference was the allocation of all usable radio frequencies according to various international radio services as opposed to allocations to states. The fixed, mobile, maritime-mobile, broadcasting, radiobeacon, air-mobile, and amateur services were all allocated frequency bands.[76]

It is interesting to note that the granting for the first time of international status to the amateur service was due in large part to long and difficult negotiations led by the United States. In 1927 amateur operations were in jeopardy because frequency allocations were needed to protect them from interference and keep them functioning. The United States recognized the work of amateurs in developing the short-wave portion of the radio spectrum and in developing reliable equipment for operation in the ultra-high frequencies (UHF). The United States also recognized the contribution of amateurs to the war effort.[77]

The 1927 conference also created the first comprehensive Table of Allocations. Adherence to this table was not made compulsory, however, as it was to be used only as a guide in assigning frequencies. This Table of Allocations, as subsequently modified and expanded, remains as the center of the ITU Radio Regulations.[78]

Another proposal to the 1927 Conference, put forth by the United States, concerned the possibilities of regional as opposed to global agreements on frequency allocations. Efforts by the United States to achieve international recognition for regional agreements, however, met with European resistance. Although the United States put forth the proposal because it felt that the time difference between the Americas and other regions negated the possibility for interference, various European nations felt that regional agreements would pit the Western Hemisphere against Europe. The issue consequently was not resolved.[79]

The issue of priority rights to radio frequencies was also addressed at the Washington Conference. A special paragraph was added to the Convention that dealt with the question of priority. It stipulated that the frequencies assigned by administrations to all new fixed, land, or broadcasting stations were "to avoid so far as possible interference with international services carried on by existing stations the frequencies of which have already been notified to the International Bureau."[80] Although the adoption of this provision at the 1927 Washington Conference failed explicitly to grant a priority right for prior use, it implicitly granted a favored status to first stations as against later stations.

Thus, the basic arrangements for regulating international radiocommunication were established in 1927. Each radio station was classified according to a defined class of use, or service, and each station was entitled to protection from harmful interference after having secured international registration. This *a posteriori* process of providing information on frequency usage, the basis for today's notification of frequency assignments to the IFRB, established the basis for current frequency management and is the primary method used for attaining rights to use radio frequencies on a noninterference basis.

One of the greatest organizational innovations created by the Washington Radiotelegraph Conference was the establishment of a radio consultative committee to study technical and related matters. Although the concept for such an advisory committee had been first put forth as early as 1919, an international radio consultative committee was not formally created until 1927.[81]

As established in 1927, the Radio Technical Consultative Committee was to meet in principle every two years. In 1932 this interval was changed to every five years and in 1938 lowered to every three years. It would be 1947 before the consultative committees would enjoy the status of permanent organs of the ITU, and henceforth meet on a regular basis.[82] More detailed coverage of the IFRB and the radio consultative committees will be provided in Chapter 4.

One further achievement of the 1927 Washington Conference was the adoption of a resolution calling for contracting administrations to examine the possibility of fusing the Telegraph and the Radiotelegraph Conventions, so as to bring both subject matters under the jurisdiction of a unified and formally constituted Union. In light of this resolution, the delegates unanimously decided to hold the next Radiotelegraph Conference simultaneously with the next meeting of the Telegraph Union in Madrid in 1932 in order to review the possibility of merging the two Unions.[83]

2.7 THE INTERNATIONAL TELECOMMUNICATION UNION

2.7.1 The 1932 Madrid Conferences

In September of 1932, both the Telegraph and the Radiotelegraph Conferences opened in Madrid. Eighty countries and 62 private companies and private and international organizations were represented at the Telegraph Conference. The Radiotelegraph Conference was attended by representatives from 65 countries and 64 private companies and private and international organizations. Both of these conferences were full diplomatic, or plenipotentiary, conferences at which delegates held powers that enabled them to negotiate and sign a treaty that would bind subscribing or ratifying governments.[84]

Ten joint meetings of the two Plenary Assemblies from the juridically separate conferences were established to consider common questions. Also established were a Joint Convention Committee and a Joint Committee on the Right to Vote. A majority vote decided all questions.[85]

The creation of a single, combined convention was the most important achievement of the Madrid Conferences. Although the majority of the delegates wanted a single convention, however, it was viewed as impossible to make all the service regulations binding on all represented countries. The United States and Canada again made it clear that they did not have the legal authority to enter into obligations that were the domain of their private operating companies. As a consequence, unanimous approval was secured by the Joint Convention Committee for an article binding contracting parties only to those regulations that they signed.[86]

The Convention of the newly created **International Telecommunication Union (ITU)** contained the broad and general principles of a more permanent nature that were considered to be common to all three telecommunication technologies — telegraph, telephone, and radio. The articles pertaining to telegraph and to telephone were generalized to the degree that no opposition was raised by the United States and Canada. Subsequently, a proposal to generalize the provisions related to radio was made by the Soviets. The attending delegates almost unanimously opposed this suggestion, stating that the universal application of provisions related to radio, such as intercommunication, noninterference, priority of distress calls, and false distress signals, were essential to the very existence of radiocommunication. After rejection of this proposal, the USSR suggested that the convention include a provision that allowed each country to make reservations to specific articles to which it felt it could not be bound. This proposal was also rejected.[87] Consequently, the Soviet delegate called for a vote on a joint convention, stating he could no longer agree to a unified convention.

The principle of a unified convention was put to a vote and passed. The delegate from the USSR then withdrew from the work of the subcommittee of the Joint Convention Committee. By the time the proposal reached full committee, however, the Soviets had succeeded in harnessing greater support. A vote in favor of inserting a formula into the convention for making reservations was passed in the Joint Convention Committee. A separate vote favoring a single convention was also passed. In line with the traditional practice of achieving consensus in the ITU, the delegates unanimously adopted a provision for inserting into the convention a way to make reservations to it.[88]

The new **International Telecommunication Convention**, which was a combination of the 1927 Radiotelegraph Convention and the 1875 Telegraph Convention, replaced the International Telegraph Union with the International Telecommunication Union. The name was chosen because, in a relatively short way, it included all three communication technologies — telegraph, telephone, and radio.[89]

Other provisions of the new Convention accomplished the following:

- A relationship was established between the consultative committees, the International Bureau, and the new Union.
- Private companies were permitted to participate in administrative conferences on an advisory basis.
- English was established as a second, official language that would be used only in conference debates and in service documents. French remained the official language of the new Union, and the costs of the English translations were borne by all interested parties.
- The practice of press censorship was curbed by making it mandatory to notify the sender of a communication if it was stopped and censored. The exception was made for communications that appeared to be dangerous to national security.
- Protection from interference was expanded for all existing stations.
- Provisions were strengthened relating to the registration of frequencies.
- The International Frequency List was officially adopted.[90]

It should be noted that the significance of the notification procedure was increased by establishing detailed rules for the format and notification procedure of frequency assignments to the Bureau. The issue of priority rights, however, was not resolved by the conference, although specific dates of first notification were to be included as part of the notification process. As a result, an increasing importance was attached to first use and first notification of a frequency.[91]

The broadcasting situation in Europe was another major concern of the Madrid Conferences. The USSR, which had not been invited to the 1927 Washington Conference, felt no obligation to adhere to the allocation table established by that conference. The Soviets, who felt bound only by the 1912 Regulations, wanted a complete reorganization of all the broadcasting bands established in 1927.[92]

Because of numerous out-of-band broadcasting operations, nine small European nations supported a revised allocation plan for broadcasting. The USSR insisted on recognition of her out-of-band operations, while the United States opposed any change in the 1927 Regulations. A solution was reached dividing up the low and medium bands into the European regions and "other regions."[93] Thus, regional agreements, initially suggested by the United States in 1927, were provided for by the new Regulations. Now the United States and Canada could continue their established operations, while a reallocation of frequencies could be achieved in Europe.

The issue of reducing interference, especially by the use of spark transmitters, was another concern at the Madrid Conference in 1932. At the 1927 Conference, the United States had succeeded in securing European agreement to ban spark transmitters by 1940, except those limited to 300 watts. While US ships were being fitted with modern, more expensive equipment, certain European nations were still equipping their ships with less costly spark transmitters. US protest was lodged and, with respect to spark transmitters, the Washington Convention was subsequently upheld. In addition, the United States was able to obtain slightly more protection for the 500 kHz distress frequency.[94]

With the signing of the Madrid Telecommunication Convention on 9 December 1932, the fusion of the Telegraph Union and the Radiotelegraph Union was accomplished.[95] The new International Telecommunication Union (ITU) came into existence as the international body that was to oversee, coordinate, and regulate the burgeoning telecommunication technologies.

2.7.2 The World War II Period

With the exception of one radio conference held in Cairo in 1938, the work of the new International Telecommunication Union came to a standstill as almost all the nations in the world became involved in the second world war. The effects of World War II on telecommunication were tremendous. The disablement and destruction of telecommunication facilities were felt by virtually every country. Although the nations of the world suffered great losses of facilities, there was still, however, significant acceleration of technological progress and development of telecommunication technologies, especially that of radio. Naturally, the simultaneous destruction and corresponding new development of telecommunication technology placed a heavy burden on post-war Union conferences to get the telecommunication networks working again.[96]

During the second world war, radio proved to be an indispensable tool, especially for military operations and propaganda broadcasts. Every branch of the military depended on radio because it was not as vulnerable to interruption of service as were telegraph and telephone. With mobile units, radio could be used anywhere and at any time. World War II served as a stimulus for many important developments in radio, many of which were quickly converted for civilian use after the war. These included the developments of the Loran Radio Navigation System, radar, and network broadcasting.[97]

With these developments in radio, however, came corresponding regulatory problems. During the war, the expansion by the military forces in radio services was largely "out-of-band," or outside the range of frequencies allocated to defined services. As a result of the expansion of services, especially in the aeronautical and high frequency broadcasting areas, allocated services in neighboring bands were threatened with encroachment. There was also no room within allocated bands for any new stations. Although the Table of Frequency Allocations of the 1938 Cairo Radio Regulations had allocated frequencies up to 200 MHz, by the end of the war the radio spectrum was extensively used up to 10,000 MHz and was partially used up to 30,000 MHz. As a result, the Table of Frequency Allocations was in urgent need of complete revision.[98]

Although a Union conference had been called in 1942 to revise the Service Regulations that had been drafted in Madrid in 1932 and revised in Cairo in 1938, it was cancelled because of World War II. The work of the ITU, however, was not completely stopped by the war. The Berne Bureau, situated on neutral territory, was able to avoid a total work stoppage. Limited wartime activities in Switzerland not only provided for the continuity of the work of the ITU during the war, but also permitted increased operations of the Union after hostilities ended.[99]

2.7.3 The 1947 Atlantic City Conferences

The first post-war ITU conferences convened in Atlantic City in 1947. The conferences held in that year were among the most important conferences in the history of the ITU and produced more changes in its structure than any other conference (see Table 2.1). At the invitation of the United States, over 70 states sent delegates to Atlantic City to participate in the three conferences convened to revise the Madrid Convention and its Service Regulations. These conferences were the Atlantic City Telecommunication Conference (Plenipotentiary), the Atlantic City World Administrative Radio Conference, and the Atlantic City International High Frequency Broadcasting Conference.[100]

Table 2.1 Early Structures and Provisions of an Evolving Union*

1865 Paris International Telegraph Conference (Plenipotentiary)
-First International Telegraph Convention and annexed Telegraph Regulations create International Telegraph Union. -Morse code designated as international telegraph alphabet.
1868 Vienna Telegraph Conference (Administrative)
-Completed original structure of International Telegraph Union by creating the International Bureau of Telegraph Administrations (Berne Bureau, secretariat).
1871 Rome Telegraph Conference (Administrative)
-Private companies given right of representation at all meetings without right to vote.
1875 St. Petersburg Telegraph Conference (Plenipotentiary)
-All elements of permanent nature placed in Convention; those subject to periodic revision go in Regulations.
1885 Berlin Telegraph Conference (Administrative)
-Short passage related to telephone added to Regulations. (Not until 1903 does telephone regulation become important part of Union work.)
1903 Preliminary Radio Conference**
-Laid basis for binding treaty on radiocommunication. -Stipulated that radio stations should operate without causing harmful interference. -Final Protocol extended various provisions of 1875 St. Petersburg Convention, including universal right to correspond by telegraphy. -Priority given communications to ships at sea. -Agreement on priority given to distress signals.
1906 Berlin Radiotelegraph Conference (Plenipotentiary)
-Several articles from 1875 Convention incorporated into convention. -Established fundamentals of modern radio regulation: • Classification of frequency bands into services allocated to specific radio uses; • *A posteriori* notification and registration of assignments.

Table 2.1 (cont'd)

-Berne Bureau designated as repository of frequency notifications and publisher of the Frequency List of Radioelectricity Stations (Berne List).
-International SOS distress signal adopted.
-*De facto* International Radiotelegraph Union emerges from convention.

1912 London Radiotelegraph Conference (Plenipotentiary)

Intercommunication between stations made obligatory.
Principles of efficiency and economy in spectrum use and management established.

1920 Washington Preliminary Conference (World Electrical Communications Conference**

-Draft convention for Universal Electrical Communications Union specifying *a priori* criteria for frequency allocations to specific countries.
-Basic structure of modern ITU proposed:

- Allocation of frequency bands among services;
- Radio consultative organ;
- Governing organ to serve in interim between conferences;
- Central bureau to conduct administrative activities.

1927 Washington Radiotelegraph Conference (Plenipotentiary)

-Scope of convention broadened to include all public radiocommunication not only public maritime correspondence.
-Principle of protection for security of correspondence established.
-Allocation to services (defined classes of use) rather than to nations.
-First comprehensive Table of Frequency Allocations created (voluntary).
-Priority via prior notification to Berne Bureau (implicit favored status).
-Radio consultative committee established.
-English as second, unofficial language added to official language of French.

1932 Madrid Conferences (Plenipotentiary and Administrative)

-Merged conventions of International Telegraph and International Radiotelegraph Unions to create International Telecommunication Union (ITU).

Table 2.1 (cont'd)

-Private participation allowed.
-Notification of censorship except for national security.
-Protection from interference expanded for all stations.
-International Frequency List adopted (dates of notification included).
-Provision for regional agreements.
-English designated as second, official language.

1947 Atlantic City Conferences (Plenipotentiary and Administrative)

-Major reorganization of the Union:

- Administrative Council created;
- Consultative Committees made permanent organs;
- General Secretariat created to replace Berne Bureau;
- IFRB created to process and examine frequency notifications.

-Regional conferences allowed to create regional agreements with radio conferences concerned with frequency allocations to be sponsored by ITU.
-Provisional Frequency Board (PFB) established to oversee engineering the spectrum.
-Rights and obligations of recorded frequencies defined to embrace new legal concepts:

- Right of international protection from harmful interference;
- Conformity with Convention and Regulations.

-Modern procedure for notification and registration established.
-Governmental priority for communications only by request.
-Absolute priority for distress calls.
-ITU became specialized agency of UN.
-Each Member country allowed one vote.
-Service Regulations obligatory upon Members, but countries given right to make reservations.
-French, English, Spanish, Russian, and Chinese designated as official languages.

---

* This table is not intended to be exhaustive, but tries to list provisions of continuing relevance or applicability to space communication.

**These conferences were not official Union conferences, but regional intergovernmental conferences.

Of course, it is not our intention here to reconstruct every issue that was before the three conferences. The work of the conferences, however, did result in an extensive reorganization of the ITU that established its present structure. Important regulatory decisions concerning frequency management were made. Therefore, because of the importance of these conferences in establishing current principles and structures related to space communications, the major changes that emanated from these conferences are briefly outlined below.

1. An Administrative Council was created to provide for continuity and to make decisions during intervals between conferences. The Administrative Council was established to control the administrative work of the Union. It was perhaps the foremost structural change at Atlantic City.[101]
2. The Consultative Committees were made permanent organs of the ITU, with provision for their assemblage on a regular basis.[102]
3. A General Secretariat was created to partially replace the existing Berne Bureau by taking over many of its functions.[103]
4. The International Frequency Registration Board (IFRB) was established to process and to examine frequency notifications. The new IFRB also partially replaced the Berne Bureau by taking over the compilation of the Master Frequency List, work that the Berne Bureau had been doing since 1928.[104]
5. The French, English, Spanish, Russian, and Chinese languages were designated as the five official languages of the Union.[105]
6. The seat of the ITU was relocated to Geneva, Switzerland, from the Swiss capital of Berne.[106]
7. The ITU became a specialized agency of the newly created United Nations organization.[107]
8. Each Member of the ITU was allowed one vote, and, thenceforth, after the 1947 Convention entered into force, only fully sovereign states could become full Members of the Union. Existing territorial or colonial members, however, were not affected. Later in 1973, this associate membership also would be abolished.[108] Previously, no definite membership article existed in the Convention that established the nature of membership or the right to vote. Under the custom of "colonial voting," however, some states had enjoyed up to six votes each.[109] The new rule, allowing for one vote per member state, laid the basis for a substantial shift in power in the ITU that resulted from the influx of developing countries — countries that began to gain their independence after World War II.

9. Provision was made in the new Convention to allow members to convene regional conferences and to enter into regional agreements. Members could also form regional organizations for the purpose of settling regional telecommunication questions. Because it was realized that tremendous conflict can arise when dealing with various portions of the radio spectrum on a regional basis, however, the stipulation was made in the Atlantic City Convention that all regional radio conferences concerned with frequency allocations were to be sponsored by the ITU.[110]
10. Finally, arrangements were made providing for the provisional operation of the newly created organs of the Union until the Convention entered into force on 1 January 1949.[111]

The Atlantic City Conferences are also important in terms of the regulatory decisions that were made in 1947. Recall that the Atlantic City Conferences were called to revise the Madrid Convention and its Radio Regulations. In light of the post-war need for revamped telecommunication regulations, the delegates to the Atlantic City Radio Conference adopted a frequency management strategy based upon the concept of an "engineered spectrum." (This concept will be more fully discussed in Chapter 4.) After three years of continuous efforts, however, the strategy proved impossible to implement.[112]

Indeed, the basic objective of the 1947 Atlantic City Radio Conference was the establishment of an engineered spectrum. This term refers to *a priori* allotments by Union conferences of the radio frequency spectrum. The concept for an engineered spectrum embraced the ideal that ITU conferences would review lists of frequency requirements submitted by the states of the world. The conferences would then adopt worldwide or regional frequency allotment plans based on these requirements. Under these plans, the specific requirements for frequency bandwidths would be internationally recognized and matched with specific frequencies used under specified technical and operating conditions.[113]

The concept for a completely engineered spectrum had to be abandoned largely because the requirements submitted by ITU Members far exceeded available frequencies.[114] Also, much of the time of the Provisional Frequency Board (PFB), the interim body set up by the Atlantic City Radio Conference to prepare a new International Frequency List of engineered frequency assignments, was spent in debate initiated by the USSR over the legality of its work. The USSR argued that the US proposed PFB was a "toy" in the hands of "certain countries" intent upon abridging the sovereignty of nations and discriminating against the Soviets. The

Soviet delegates argued that any new frequency list would be "prejudicial" to their interests, and the delegates reserved the right to use all the frequencies that had been registered in the Berne List and any that might be registered later. The Soviets insisted that any frequency assignments be based on the 1939 Berne List, keeping the established notification date.[115]

There were more than 2,000 frequencies notified by the Soviets in the Berne List. The notifications were on every 10 kHz and in many cases on every 5 kHz. The Soviet delegate argued that the drafting of a new International Frequency List was an attempt, particularly by the United States, to redistribute radio frequencies to satisfy their own inflated requirements. In essence, the USSR did not want to relinquish the favored legal status of the great many frequency notifications it had registered to the Berne Bureau prior to World War II.[116]

As a result of this international conflict in the PFB and of the excessive requests for frequency requirements, the task of drawing up a new frequency list was never accomplished and had to be given up. In its wake, however, was left a regulatory system that serves as the basis for the present legal regime for frequency management. This is especially true concerning the principles that established the legal status of recorded frequency assignments.[117]

The new legal principles set forth in Atlantic City were as follows.[118]

- The rights and obligations concerning recorded frequency assignments involved two new legal concepts: first, the right of international protection from harmful interference, and, second, the principle of conformity with the Convention and Regulations. Additionally, the prewar significance given to earlier use and notification was retained in certain circumstances.
- The IFRB was created and given various duties, such as the legal and technical examination of new frequency notifications. The Board was empowered to issue findings on the notifications with respect to their conformity with the Convention and Regulations and with respect to their potential for causing harmful interference to other stations.
- An elaborate procedure was set up for the notification and registration of frequency assignments.

Another regulatory decision made in Atlantic City concerned the prohibition of type-B emissions from spark transmitters. Prior conferences had failed in their attempts to prohibit such emissions completely because of intense opposition from countries and companies that held large maritime interests. The Atlantic City Radio Conference established a provision in the Radio Regulations that forbade all stations from using type-B emissions, with the exception of emergency installations in ships and on lifeboats or other survival equipment.[119]

Also, priority was given over all other communications to government telegrams and telephone calls. This was done only if so requested by the sender, however. Prior to 1947, all government communications were automatically given priority in transmission. The practical effect of the new provision that expressly required a request for priority was to set back the old principle that telecommunication systems were established primarily for the needs of the state. In addition, absolute priority for distress calls was extended beyond radio messages to telegraph and telephone communications.[120]

Another provision in the Atlantic City Convention made the Service Regulations obligatory upon ITU Members. At the same time, however, countries were given the right to make specific reservations to the different sets of Regulations that were annexed to the Convention.[121]

2.8 CONCLUSION

The principles established in 1947 have historical origins at least as far back as the 1875 St. Petersburg Convention and the 1906 Radiotelegraph Convention. In 1906 classification of frequency bands into services allocated to specific radio uses was begun. Also begun in 1906 was *a posteriori* notification to and recordation by the Berne Bureau. The 1912 Conference followed the lead of the 1906 Conference by becoming the first to give serious consideration to the principles of efficiency and economy in radio spectrum use and management.

The 1927 Radiotelegraph Conference recognized the principle of protection of the security of private correspondence. The Conference also created the first Table of Frequency Allocations and established, at least implicitly, the concept that priority is to be accorded first notification to the Berne Bureau. The 1932 International Telecommunication Convention followed by establishing the International Telecommunication Union, adopting provisions for regional agreements, allowing private participation in the ITU, requiring notification by the state of censorship, and adopting an International Frequency List with dates of first notification of service included.

The pivotal 1947 Telecommunication Convention established the current structure of the ITU, with the Administrative Council and the Consultative Committees made permanent organs, and with the General Secretariat and the International Frequency Registration Board established to replace the Berne Bureau. The 1947 Atlantic City Convention also established a legal right of international protection from harmful interference, established an obligation for states to be in conformity with the Convention and its annexed Radio Regulations, and completed the establishment of the modern *a posteriori* procedure for notification and recordation of the uses of radio that would be afforded international protection from harmful interference.

Legal principles established in 1947 form the basis of a regulatory regime for space communication that exists to this day. Although other regulations and rules were set forth by the Atlantic City Convention, only those having a heavy bearing upon space communication have been discussed here. Suffice it to say that the changes made by the new Convention were broad in scope and have had a lasting impact.

Based upon this historical overview of the activities of the nations of the world in forming international organizations to coordinate and to regulate telecommunication, we now turn to an examination of the structure and the workings of the ITU in terms of its universal, technical impact on space radiocommunication.

REFERENCES

1. Codding, George A. Jr., *The International Telecommunication Union,* Leiden, Netherlands: E.J. Brill, 1952, p. 6.
2. DeFleur, M.L. and S. Ball-Rokeach, *Theories of Mass Communication,* (4th. Edition), New York: Longman, 1982, p. 71.
3. Codding, *supra* note 1, at 7.
4. Feldman, Mildred L.B., *The Role of the United States in the International Telecommunication Union and Pre-ITU Conferences,* Baton Rouge, LA: Mildred L. Bos Feldman, 1976, p. 4.
5. *Ibid.*
6. *Ibid.*
7. Codding, George A. Jr., and Anthony M. Rutkowski, *The International Telecommunication Union in a Changing World,* Dedham, MA: Artech House, 1982, p. 5.
8. Feldman, *supra* note 4, at 4.
9. Leive, David M., *International Telecommunications and International Law: The Regulation of the Radio Spectrum,* Dobbs Ferry, NY: Oceana Publications, 1970, p. 9.

10. Codding, *supra* note 1, at 24.
11. Feldman, *supra* note 4, at 5.
12. Codding, *supra* note 1, at 27.
13. *Ibid.*, p. 26.
14. *Ibid.*, pp. 26, 51.
15. *Ibid.*, p. 29.
16. Codding and Rutkowski, *supra* note 7, at 16.
17. *Ibid.*, p. 9.
18. Codding, *supra* note 1, at 32.
19. Codding and Rutkowski, *supra* note 7, at 12.
20. Codding, *supra* note 1, at 84.
21. Feldman, *supra* note 4, at 58.
22. *Ibid.*, p. 55.
23. *Ibid.*, p. 5.
24. Codding and Rutkowski, *supra* note 7, at 12.
25. Feldman, *supra* note 4, at 57.
26. Codding, *supra* note 1, at 87.
27. *Ibid.*
28. Feldman, *supra* note 4, at 25.
29. Codding, *supra* note 1, at 87.
30. *Ibid.*, pp. 51, 81.
31. Codding and Rutkowski, *supra* note 7, at 3.
32. *Ibid.*, p. 13.
33. *Ibid.*
34. *Ibid.*, p. 261.
35. Leive, *supra* note 9, at 57.
36. Codding and Rutkowski, *supra* note 7, at 13.
37. Feldman, *supra* note 4, at 28.
38. Codding and Rutkowski, *supra* note 7, at 16.
39. Codding, *supra* note 1, at 127.
40. *Ibid.*, p. 111.
41. *Ibid.*, p. 127.
42. *Ibid.*, p. 128.
43. Codding and Rutkowski, *supra* note 7, at 36, 53.
44. *Ibid.*, p. 53.
45. Codding, *supra* note 1, at 129.
46. Codding and Rutkowski, *supra* note 7, at 203, 212.
47. *Ibid.*, p. 14.
48. Codding, *supra* note 1, at 103.
49. Feldman, *supra* note 4, at 33.
50. Service Regulations Affixed to the *International Radiotelegraph Convention, London, 1912,* Art. 7, quoted in Codding and Rutkowski, *supra* note 7, at 255.

51. Codding, *supra* note 1, at 109.
52. Feldman, *supra* note 4, at 8.
53. *Ibid.*, p. 9.
54. *Ibid.*, p. 34.
55. Codding and Rutkowski, *supra* note 7, at 255.
56. *Ibid.*, p. 86.
57. Leive, *supra* note 9, at 43.
58. Codding and Rutkowski, *supra* note 7, at 256.
59. Feldman, *supra* note 4, at 11.
60. *Ibid.*, p. 12.
61. *Ibid.*, p. 10.
62. *Ibid.*, p. 13.
63. *Ibid.*, p. 14.
64. Glazer, J. Henry, "INFELIX ITU - The Need for Space-Age Revisions to the International Telecommuncation Convention," *Federal Bar Journal,* Vol. 23, 1963, pp. 1-36, at 19.
65. An *allocation* is defined in the ITU Radio Regulations as an "[e]ntry in the Table of Frequency Allocations of a given frequency band for the purpose of its use by one or more terrestrial or space radiocommunication services or the radio astronomy service under specified conditions. This term shall also be applied to the frequency band concerned." An *assignment* is "[a]uthorization given by an administration for a radio station to use a radio frequency or radio frequency channel under specified conditions." An *allotment* is an "[e]ntry of a designated frequency channel in an agreed plan, adopted by a competent conference for use by one or more administrations for a terrestrial or space radiocommunication service in one or more identified countries or geographical areas and under specified conditions."
66. Codding and Rutkowski, *supra* note 7, at 15.
67. *Ibid.*, pp. 15, 264.
68. Codding, *supra* note 1, at 111.
69. Feldman, *supra* note 4, at 35.
70. Codding, *supra* note 1, at 116.
71. Feldman, *supra* note 4, at 65.
72. Codding, *supra* note 1, at 118.
73. Feldman, *supra* note 4, at 41.
74. Codding, *supra* note 1, at 119, 121.
75. *Ibid.*, p. 124.
76. Codding and Rutkowski, *supra* note 7, at 257.
77. Feldman, *supra* note 4, at 37, 44.
78. Codding and Rutkowski, *supra* note 7, at 257.
79. Feldman, *supra* note 4, at 43.

80. Leive, *supra* note 9, at 47.
81. Codding and Rutkowski, *supra* note 7, at 86.
82. Codding, *supra* note 1, at 122, 151, 169, 298.
83. *Ibid.*, p. 130.
84. *Ibid.*, p. 131.
85. *Ibid.*
86. *Ibid.*, p. 137.
87. *Ibid.*, p. 138.
88. *Ibid.*
89. *Ibid.*, p. 140.
90. *Ibid.*, pp. 140-150.
91. Leive, *supra* note 9, at 48.
92. Feldman, *supra* note 4, at 76.
93. Codding, *supra* note 1, at 149.
94. Feldman, *supra* note 4, at 78.
95. *International Telecommunication Convention, Madrid, 1932.*
96. Codding, *supra* note 1, at 181.
97. *Ibid.*, p. 182.
98. *Ibid.*, p. 185.
99. *Ibid.*, p. 193.
100. *Ibid.*, p. 223.
101. *Ibid.*, p. 286.
102. *Ibid.*, p. 271.
103. *Ibid.*, p. 291.
104. *Ibid.*, p. 241.
105. *Ibid.*, p. 304.
106. *Ibid.*, p. 308.
107. *Ibid.*, p. 320.
108. *Ibid.*, pp. 205, 279.
109. Feldman, *supra* note 4, at 41.
110. Codding, *supra* note 1, at 311.
111. *Ibid.*, p. 327.
112. Leive, *supra* note 9, at 56.
113. *Ibid.*
114. *Ibid.*, p. 68.
115. Codding, *supra* note 1, at 355, 358.
116. *Ibid.*, p. 353.
117. Leive, *supra* note 9, at 56.
118. *Ibid.*, p. 57.
119. Codding, *supra* note 1, at 264.
120. *Ibid.*, p. 322.
121. *Ibid.*, p. 325.

PART II

STRUCTURE OF THE SPACE COMMUNICATION LAW-MAKING PROCESS IN THE INTERNATIONAL TELECOMMUNICATION UNION

Chapter 3
The Law-Making Conferences of the International Telecommunication Union and Space Communication

3.1 INTRODUCTION

Satellite communication has emerged over the last twenty-five years to become a central component of modern communication systems, both domestic and international. Communication via satellite represents a significant economic activity on the order of $2 billion per year, with projections reaching $10 billion by 1990.[1]

Yet in order to understand the law and regulation of satellite communication it is necessary to be familiar with the specialized international organization responsible for facilitating and administering the regulation by treaty of international space communication. As we have seen, that organization is the International Telecommunication Union (ITU), which is charged with coordinating and regulating the complex, international activities that collectively constitute a global telecommunication network.

In 1987, 160 nation-states belonged to the ITU. These states, bound by agreement to all the pertinent provisions of the International Telecommunication Convention and its annexed Radio Regulations, Telegraph Regulations and Telephone Regulations, have the right to participate and to vote at all conferences held by the ITU.

3.1.1 Purposes of the Union

The original *raison d'être* of the ITU, bequeathed by its predecessor, the International Telegraph Union, was to set international telecommunication standards. The purposes of the ITU today extend beyond setting standards and include the following principles:

- "to maintain and extend international cooperation between all Members of the Union for the improvement and rational use of telecommunications of all kinds, as well as to promote and to offer technical assistance to developing countries in the field of telecommunications;
- to promote the development of technical facilities and their most efficient operation . . . ; [and]
- to harmonize the actions of nations in the attainment of those ends."[2]

The ITU's most significant function is the regulation of the world's international telecommunication networks in order to maintain an efficient, worldwide system. To this end the Union shall:

- "effect allocation of the radio frequency spectrum and registration of radio frequency assignments in order to avoid harmful interference between radio stations of different countries; [and]
- coordinate efforts to eliminate harmful interference between radio stations of different countries and to improve the use made of the radio frequency spectrum. . . . "[3]

In addition to the central role it now plays in management of worldwide use of the radio spectrum, other important functions of the ITU are to:

- "foster international cooperation in the delivery of technical assistance to the developing countries and the creation, development and improvement of telecommunication equipment and networks in developing countries by every means at its disposal . . . ;
- coordinate efforts with a view to harmonizing the development of telecommunication facilities, notably those using space techniques . . . ;
- foster collaboration among its Members with a view to the establishment of rates at levels as low as possible consistent with an efficient service . . .;
- promote the adoption of measures for ensuring the safety of life through the cooperation of telecommunication services; [and]
- undertake studies, make regulations, adopt resolutions, formulate recommendations and opinions, and collect and publish information concerning telecommunication matters."[4]

These provisions of the 1982 International Telecommunication Convention, which is international law binding upon all signatory states, define the mission of the ITU — the maintenance of an efficient, worldwide telecommunication network through the constant upgrading of the technologies and regulations facilitating network operation.

3.1.2 Structure of the Union

To accomplish the mission of the ITU, the modern organizational structure of the ITU consists of the following organs (also see Table 3.1):

- The Plenipotentiary Conference;
- Administrative Conferences;
- The Administrative Council; and
- The Permanent Organs.

The Plenipotentiary and Administrative Conferences, as well as the Administrative Council could be described as Periodic Organs of the ITU, in that they are not always in operation or in session, but rather are called into session periodically according to procedures established in the International Telecommunication Convention. The Permanent Organs of the ITU are in continual operation, with permanent staffs, and consist of the following:

- The General Secretariat;
- The International Frequency Registration Board (IFRB);
- The International Radio Consultative Committee (CCIR); and
- The International Telegraph and Telephone Consultative Committee (CCITT).[5]

Each structural component of the Union serves a purpose, performing designated tasks and discharging specific responsibilities. International law is made in the Plenipotentiary and Administrative Conferences. The focus of this chapter is the structure surrounding the law-making conferences of the Union, as well as the primary activities and responsibilities of the conferences themselves.

3.2 THE PLENIPOTENTIARY CONFERENCE

The Plenipotentiary Conference is the supreme organ of the ITU. "Plenipotentiary" denotes full power, and this is true of the Plenipotentiary Conference, in that it has not only regulatory authority but also constitutional authority. The plenipotentiary has the authority to change the constitutional document of the ITU — the International Telecommunication Convention.

As is characteristic of international organizations, decisions of importance are made at plenipotentiary conferences composed of delegates from member countries. In principle the Plenipotentiary Conference of

Table 3.1 Structure of the ITU

Source: Codding, George A., Jr., and Anthony M. Rutkowski, *The International Telecommunication Union in a Changing World,* Dedham, MA: Artech House, 1982. Reprinted with permission.

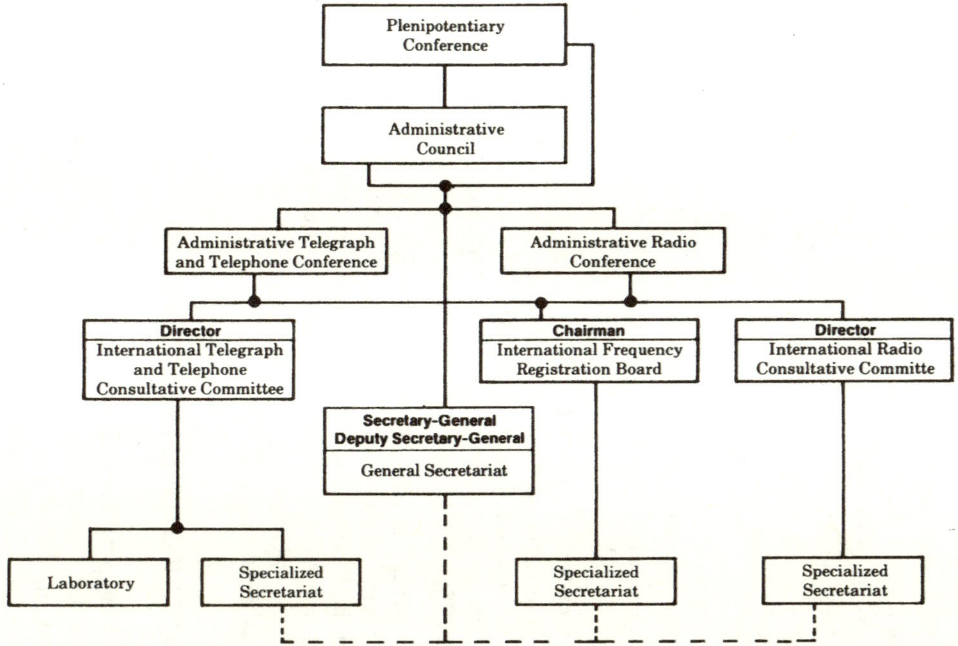

Note: The ITU Convention designates the CCIR, CCITT, IFRB, and the General Secretariat as the permanent organs of the ITU. The ITU also has a coordination committee (not shown), made up of the Secretary General, Deputy Secretary General, Director of the CCIR, Director of the CCITT, and Chairman of the IFRB, whose major task it is to "assist and advise" the Secretary General. See ITU, ***International** Telecommunication Convention,* Arts. 5 and 12, Nairobi, 1982.

the ITU is to be convened every five years.[6] However, the location and date of a conference can be established by the previous plenipotentiary or can be fixed by the Administrative Council with the approval of a majority of Members of the Union.[7]

3.2.1 The International Telecommunication Convention

The ITU's basic, constitutional document — the International Telecommunication Convention — is an instrument that must be ratified by the Member countries before it is binding upon them. The Convention gives the ITU its legal existence, establishes its composition, purposes, structure, functions, general provisions relating to telecommunication and specific provisions related to radiocommunication. The Convention also sets forth the relations of the ITU with the United Nations and other international organizations, and establishes administrative regulations and rules of procedure for conferences and other meetings.

Other provisions of the Convention deal with financial matters, arbitration procedures and the official languages of the ITU. As of 1987 there were six official languages: Arabic, Chinese, English, French, Russian, and Spanish. The final documents of the plenipotentiary and administrative conferences are drawn up in the official languages. All other documents of these conferences are issued in the working languages of the ITU: English, Spanish, and French, with French prevailing in case of dispute.[8]

3.2.1.1 Structure of the Convention

Following the basic provisions of the Convention are the signatures of the representatives of those states whose delegations have acceded to the treaty. Annexed to the treaty document is a list of the signatory countries (Annex 1), definitions of certain terms used in the Convention and in the Administrative Regulations (Annex 2), and the 1947 agreement of affiliation between the United Nations and the ITU (Annex 3). The treaty provisions and signatures are then followed by a Final Protocol.[9]

As permitted by the rules of international law, any State may take a Reservation to a treaty. It is in the Final Protocol, which forms part of the Final Acts of the particular plenipotentiary conference, where Member countries state their Reservations. The **Reservations** are listed in the chronological order of their deposit and constitute reservations by particular Member countries regarding certain aspects of the Convention, its attached

Annexes and Protocols. Appended to the Convention are additional Protocols, Resolutions, Recommendations, and Opinions.

Resolutions generally take the form of directives to other bodies of the ITU. Most of the 75 resolutions contained in the 1982 Nairobi Convention were carry-overs from previous conventions, although 26 were new.[10] Of all of the 75 resolutions in the Nairobi Convention, about one-third concerned the developing countries, with 18 resolutions dealing with technical cooperation.

Recommendations generally contain suggestions to Member countries and to the United Nations. One recommendation, concerning the unrestricted transmission of news, was appended to the Nairobi Convention.

Opinions appended to the treaty preserve the views of the delegates to that plenipotentiary conference. There were three opinions attached to the Nairobi Convention. These dealt with the avoidance of fiscal taxes on any international telecommunication, favorable treatment for developing countries regarding telecommunication, and the need for international telecommunication exhibitions.[11]

The current Convention, enacted at the 1982 Plenipotentiary Conference held in Nairobi, contained an Optional Additional Protocol signed by 91 Member countries. This Protocol deals with the compulsory settlement of disputes related to disagreement concerning the interpretation or application of the Convention or of its Administrative Regulations, such as the Radio Regulations. In the event of dispute and at the request of one of the parties to the dispute, the signatory States have agreed to compulsory arbitration.[12]

3.2.1.2 Major Provisions of the Convention

In addition to its authority to revise the Convention, the Plenipotentiary Conference has been charged by the 1982 Convention with several other important responsibilities. They include:

- Determination of general policies for fulfilling the purposes of the Union;
- Consideration of the report by the Administrative Council on the activities since the previous Plenipotentiary of all the organs of the Union;
- Establishment of the budget of the Union, setting the fiscal limit for the expenses of the Union until the next Plenipotentiary Conference;
- Establishment of the basic salaries, salary scales and the allowance and pension system for all the officials of the Union;
- Examination and, where appropriate, approval of the accounts of the Union;

- Election of the Members of the Union to serve on the Administrative Council, the Secretary-General, the Deputy Secretary-General, the members of the IFRB, and the directors of the International Consultative Committees; and
- Conclusion or revision, when necessary, of any agreement between the ITU and other international organizations.[13]

3.2.2 Structure and Procedure of the Plenipotentiary Conference

The formal structure of the Plenipotentiary Conference consists of delegates from Member countries who, along with their professional staff and non-voting RPOAs and SIOs, work together in various committees, sub-committees, and working groups. An informal structure also exists where unofficial consultations between groups of delegates often work to expedite much of the work that ends up in formal committees.

If there is a host government for the Plenipotentiary Conference, the chairman of the conference, who is elected at the first full, or "Plenary," meeting, is generally nominated by that government. If there is no inviting government, the chairman is chosen from proposals formulated at the Heads of Delegations meeting, which convenes just prior to the opening of the conference, and also prepares a list of candidates for other conference officers. In addition to the chairman, the first Plenary Meeting selects the conference vice chairmen, sets up the conference committees, and selects the chairmen and vice chairmen of the respective committees.[14]

Each plenipotentiary has a Steering Committee, a Credentials Committee, a Budget Control Committee, and an Editorial Committee. The Steering Committee is made up of the chairman and vice chairmen of the conference and of the chairmen and vice chairmen of all the standing committees of the conference, with the conference chair also serving as chair of this committee. The Steering Committee ensures the smooth working of the conference by coordinating all pertinent matters and by planning and ordering the meetings.[15]

The Credentials Committee is responsible for verifying the credentials of delegations to the conference. The Budget Control Committee not only oversees the finances of the conference, but also alerts the Plenary Meeting if the approved budget may be exceeded, issues a report of the total expenditures of the conference and estimates the costs that may be incurred by the execution of the decisions taken by the conference.[16]

The Editorial Committee, which traditionally has had a French chairman, has two vice chairmen, one from an English-speaking country and one from a Spanish-speaking country.[17] This committee is charged with reconciling the texts prepared in the various committees and editing their form without changing their substance. The texts are then submitted to the Plenary Meeting, which approves or refers them back to the appropriate committee.[18]

Plenipotentiary conferences also have committees devoted to proposals for changes in the Union's structure and to matters related to the general regulations for telecommunication and the specific regulations for radiocommunication. Other committees might deal with matters related to the ITU's finances, to relations with the United Nations and its development assistance programs, and to questions of personnel in the Secretariat.

Reaching consensus on important issues is in the tradition of the ITU. Still, more and more often, voting does occur, with all decisions made by a simple majority of those present and voting. The exception is in cases concerning the admission of new Members to the Union. In these instances, a two-thirds approval by the Members is needed. In computing a majority, abstentions are not taken into account, but for a valid vote to be taken at a Plenary Meeting, a quorum is needed. A quorum is defined as more than one-half of the delegations accredited to the conference having the right to vote, present or represented at the meeting.[19] Since 1959, however, Member countries that have not ratified the Convention within two years of its execution lose their right to vote in conferences, meetings and consultations.

3.3 ADMINISTRATIVE SUPPORT FOR UNION CONFERENCES

3.3.1 The General Secretariat

In support of Union conferences, it is the responsibility of the General Secretariat to gather together all the proposals submitted by Member countries and to correlate the proposals according to the part of the Convention to which they relate. The Secretariat prepares recommendations for the meeting of the Heads of Delegations where the agenda for the first Plenary Meeting of the Plenipotentiary Conference is set and where proposals for the organization, chairmanships and vice chairmanships of the conference and its working committees are developed.[20]

The Secretariat arranges for the meeting places, produces the minutes of the meetings and publishes the final report of the conference. It provides for translation and interpretation services, organizes the distribution of documents and gives advice to the chairman and vice chairmen of the conference, its committees and its working groups.

As is the case in every international organization, the Secretariat is the foundation upon which the whole organization rests, in that it is a permanent organ that provides day-to-day continuity. The General Secretariat is headquartered in Geneva, Switzerland, and, as noted in Chapter 2, is a direct descendent of the old Berne Bureau.

Recall that although the Berne Bureau had been operated by the Swiss, the 1947 Atlantic City Telecommunication Conference decided to internationalize the Secretariat, placing the election of its director into the hands of the Members of the Union. The underlying philosophy was to have personnel from many nations working together as a sort of "international civil service" based on the model of the United Nations.[21]

The duties charged to the General Secretariat by the International Telecommunication Convention primarily fall within four general categories — administrative, staff, support work for conferences and meetings, and information distribution.[22] To accomplish these functions the Secretariat, headed by the Secretary-General and assisted by a Deputy Secretary-General, is organized into six departments — personnel, finance, conferences and common services, computer, external relations, technical cooperation, and archives.[23] (See Table 3.2.)

3.3.2 The Administrative Council

The Administrative Council is the official agent of the Plenipotentiary Conference in the interval between plenipotentiaries. While it does not possess the full authority of a plenipotentiary, there are many matters that arise over the several years interval between plenipotentiaries that require the attention of a kind of "executive committee" of the ITU. The Administrative Council is that executive committee, with its nature and the scope of its authority spelled out in the International Telecommunication Convention.

The Administrative Council, which meets each year in Geneva, is a periodic organ composed of delegates from 41 Member countries 1987 who are elected by the Plenipotentiary Conference "with due regard to the need for equitable distribution of the seats on the Council among all regions of the world."[24] At the beginning of each annual session, the Council elects its own chairman and vice chairman with the principle of rotation of officers observed in such selections. The Secretary-General serves as secretary of the Council.[25]

The Administrative Council performs many duties associated with plenipotentiary and administrative conferences, including setting the agenda for administrative conferences. The Council performs duties assigned to it by the previous Plenipotentiary Conference and exercises financial control over the permanent organs. In addition the Council is empowered to:

Table 3.2 The General Secretariat of the ITU
Source: Codding, George A., Jr., and Anthony M. Rutkowski, *The International Telecommunication Union in a Changing World*, Dedham, MA: Artech House, 1982. Reprinted with permission.

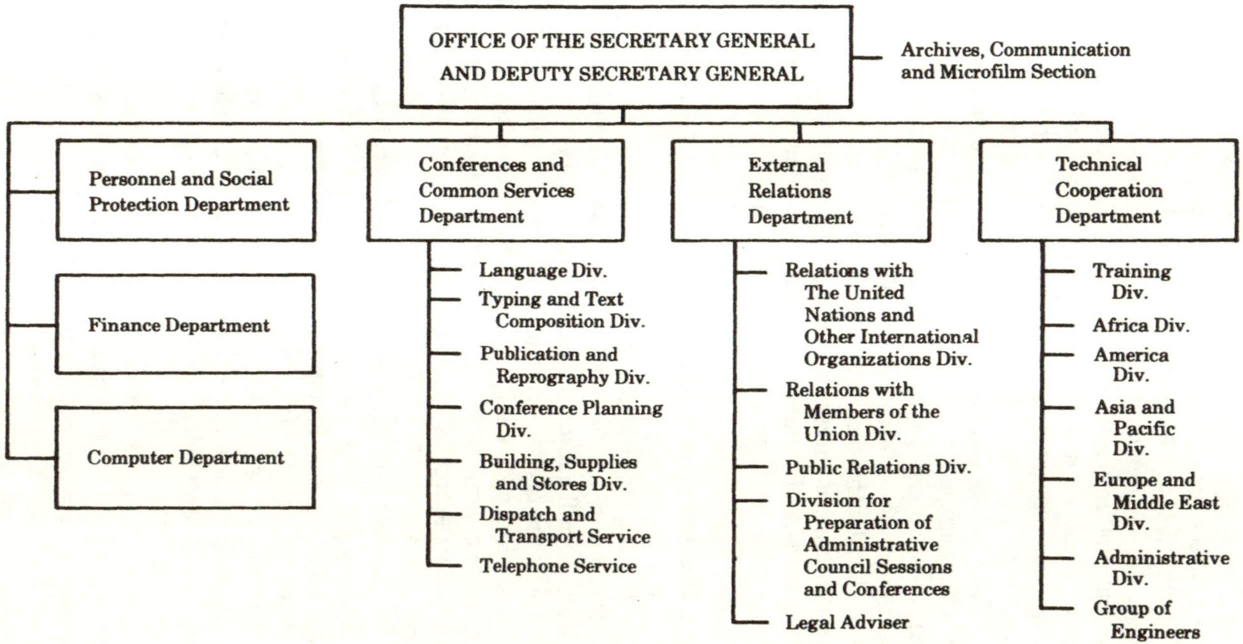

- decide on proposals for operational changes within the Union's permanent organs;
- establish and maintain personnel and salary standards;
- review and coordinate the ongoing work, programs, and progress of the permanent organs;
- appoint successors in case of vacancies to the positions of Secretary-General, Deputy Secretary-General, directors of the consultative committees, and members of the IFRB;
- report to the Plenipotentiary Conference on the activities of all the organs of the Union since the previous plenipotentiary; and
- send summary records of the activities of the Council after each of its sessions to the Members of the Union.[26]

3.3.3 The Coordination Committee

The Coordination Committee is comprised of the Secretary-General, who serves as its chairman, the Deputy Secretary-General, the directors of the consultative committees, and the chairman and vice chairman of the IFRB. The Committee is to meet at least once a month or at the request of two of its members, and is to issue a report upon request of the Administrative Council.[27]

The concept of the Coordination Committee is to fashion a bridge between the autonomous permanent organs of the ITU. Embraced within the structural reorganization of the ITU in 1947 was a sort of federal or confederated structure, involving a kind of separation of powers, in that the IFRB and the consultative committees were given substantial autonomy (see Chapter 4). In addition, these organs contained specialists who might not have been found in the General Secretariat.

Duties charged to the Coordination Committee include giving advice and practical assistance to the Secretary-General on all matters affecting more than one permanent organ, and on matters related to public relations and external relations, including coordination with the United Nations and other international organizations that have interests and responsibilities related to telecommunication. The Committee is also to consider any matters referred to it by the Administrative Council that are within its jurisdiction as defined by the Convention.[28]

3.4 ADMINISTRATIVE CONFERENCES

There are two types of ITU administrative conference, the world conference and the regional conference. These conferences can be called to consider specific telecommunication matters dealing either with telegraph and telephone or with radiocommunication. Matters of space radiocommunication are dealt with in world administrative radio conferences (WARCs) or in regional administrative radio conferences (RARCs).

Administrative radio conferences are convened by either a decision of a plenipotentiary conference, or by a recommendation of a previous world administrative radio conference with the approval of the Administrative Council, by a specific proposal of the Administrative Council, or at the request to the Secretary-General of at least one-fourth of the Members of the ITU. The date, place, and budget are set by the Administrative Council with a concurrence of a majority of Members who have replied within time limits set by the Council. If the number of replies does not exceed one-half of the Members consulted, another consultation occurs, which is decisive regardless of the number of votes cast. If the decision to convene is made by a plenipotentiary conference, this conference also sets the date, budget, and place.[29]

Regional administrative radio conferences are called by the same process, with the additional provision that a previous regional administrative radio conference can call the meeting. The votes necessary to convene a regional conference are predicated on approval of one-fourth of the Members from the region involved. The International Telecommunication Convention designates three geographical regions eligible to convene regional conferences (see Figure 3.1). The regions are:

- Region 1 — Europe, Africa, the Middle East, the USSR, and Mongolia;
- Region 2 — The Americas, the Caribbean, and Greenland; and
- Region 3 — Asia and the Pacific Basin other than Hawaii.[30]

3.4.1 Structure and Procedure of Administrative Conferences

The agenda of an administrative radio conference is established by the Administrative Council with the approval of a majority of Members for a world radio conference, or, in the case of a regional radio conference, by a majority of Members from the particular region. Any question that a plenipotentiary conference has placed on the agenda must be included, with conference discussion limited to actual agenda items.[31]

The agenda of a world administrative radio conference may include either partial or complete revision of the ITU Radio Regulations. The agenda for a regional administrative radio conference can only deal with specific telecommunication questions of a regional nature.[32] The agenda for either type of conference can be changed under the same procedures as govern date and location changes.[33]

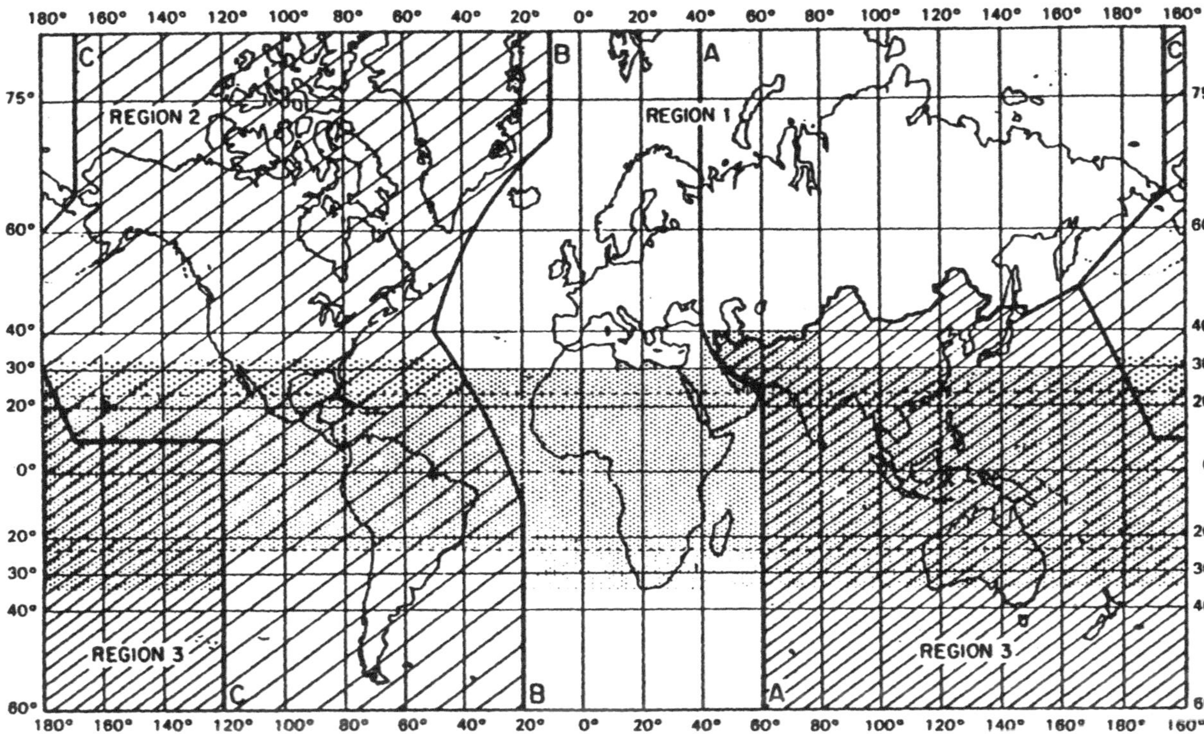

The shaded part represents the Tropical Zone as defined in Nos. 406 to 410 and 411.

Figure 3.1 ITU Regions of the World
Source: ITU, *Radio Regulations*, Art. 8, Geneva, 1982.

As regards preparatory work for an administrative conference, either a plenipotentiary conference or the Administrative Council may call for a preparatory session, which would draft and issue to the conference a report on the technical aspects of its agenda. At the invitation of a plenipotentiary, the Administrative Council, or an administrative radio conference, the International Radio Consultative Committee (CCIR) can also convene a preparatory meeting for a radio conference. The document generated by this preparatory meeting can also be submitted to the radio conference.[34] In addition, the International Frequency Registration Board (IFRB) generally provides technical assistance in making preparations for radio conferences.[35]

The general organization and election process of an administrative conference and its committees are similar to those of a plenipotentiary conference. Outside of the Steering, Credentials, Budget Control, and Editorial Committees, the needs of each conference dictate the types of other committees that are set up at an administrative conference, just as particular needs dictate this organizational structure at a plenipotentiary conference. Most conferences, however, have a technical regulations committee, a frequency allocations committee, and a committee on regulatory procedures. Committees are typically also given a number, with, for example, Steering as Committee One, Credentials as Committee Two, Budget as Committee Three, *et cetera*. Numbers of other committees may vary, although Committee Four usually considers technical regulations, Committee Five deals with allocations, and Committee Six treats regulatory procedures.

There have been three **general** world administrative radio conferences (GWARCs) since World War II with the broad jurisdiction to consider complete revision of the Radio Regulations. The GWARCs occurred in Atlantic City in 1947, in Geneva in 1959, and there again in 1979. More than 15 WARCs with more specialized or limited jurisdiction have been held since 1947, and although no regional conferences related to telegraph and telephone matters were necessary during this time, about a dozen regional administrative conferences related to radio were held.[36]

3.4.2 The International Radio Regulations

The Radio Regulations, the final product of an administrative radio conference, complement and are annexed to the International Telecommunication Convention. The Radio Regulations are binding upon all Members of the Union. Because each country has the sovereign right to regulate its own telecommunication environment, however, states can make specific reservations to the different Radio Regulations just as they can to the Convention.

The Radio Regulations are divided into two parts. Part A, the General Provisions, includes provisions that deal with terminology, measures against interference, and administrative provisions for radio stations. The regulations for frequency allocations to the various radio services and the procedures for coordination, notification, and registration to the IFRB of these frequencies are also in Part A of the Radio Regulations. As of 1987, 37 radiocommunication services were differentiated by the ITU, 17 of which are recognized by the Table of Frequency Allocations as being space radiocommunication services.[37]

Two chapters of Part A of the Radio Regulations, Chapter 3 on frequency allocations and Chapter 4 on IFRB recordation procedures, stand as the heart of and the most contentious aspects of the Radio Regulations. In Chapter 3, Article 8 of the General Provisions of the Radio Regulations is the Table of Frequency Allocations, which specifies the allocation of radio frequency bands from 9 kHz to 400 GHz to the various radio services throughout the world. An attempt is made in the allocation process to require the same worldwide usage for frequencies that have global propagation characteristics. Sharing among different users of the same frequency bands, however, either on a worldwide or on a regional basis, often occurs. For instance, the highly desirable 12 GHz band is shared by various services including the fixed, fixed-satellite, broadcasting, and broadcasting-satellite services.[38]

Individual states and groups of states can also record a "footnote" to the Frequency Allocation Table. A footnote indicates an exception, limitation, or condition of usage for the frequency band other than that stipulated in the Table. These footnotes have greatly increased over the years and must be considered if efficient and effective frequency management is to ensue. In practice, they greatly complicate efficient and equitable mediation of the demands for use of the radio spectrum (see Table 3.3).

Also located in Part A of the Radio Regulations, in Chapter 4, are the procedures that have come to be called priority rights-vesting mechanisms. These mechanisms refer to the way a state secures the right of frequency use free from interference by stations of other states. This chapter of the Radio Regulations indicates the method to be used by states for notification and registration to the IFRB of radio frequency assignments. Chapter 4 also indicates the coordination procedure required to be carried out among nations before a new satellite communication system is put into service. The registration procedure used by the IFRB to afford recognition to these notifications is also established here. This is discussed in more detail in Chapter 4.

In Part B of the Radio Regulations are technical provisions relating to groups of services and to specific services and stations. The provisions of this part of the Radio Regulations contain various rules and technical

Table 3.3 A Portion of the Table of Frequency Allocations Showing Allocations to Services and Partial Listing of Footnotes
Source: ITU, *Radio Regulations,* Art. 8, Geneva, 1982.

GHz
11.7 — 12.75

<table>
<tr><th colspan="3">Allocation to Services</th></tr>
<tr><th>Region 1</th><th>Region 2</th><th>Region 3</th></tr>
<tr>
<td rowspan="4">11.7 — 12.5
FIXED
BROADCASTING
BROADCASTING-SATELLITE
Mobile except aeronautical mobile

838 840</td>
<td>11.7 — 12.1
FIXED 837
FIXED-SATELLITE (space-to-Earth)
Mobile except aeronautical mobile

836 839 840</td>
<td rowspan="2">11.7 — 12.2
FIXED
MOBILE except aeronautical mobile
BROADCASTING
BROADCASTING-SATELLITE

838 840</td>
</tr>
<tr>
<td rowspan="2">12.1 — 12.3
FIXED 837
FIXED-SATELLITE (space-to-Earth)
MOBILE except aeronautical mobile
BROADCASTING
BROADCASTING-SATELLITE

839 840 841
842 843 844</td>
</tr>
<tr>
<td rowspan="2">12.2 — 12.5
FIXED
MOBILE except aeronautical mobile
BROADCASTING

838 840 845</td>
</tr>
<tr>
<td rowspan="2">12.3 — 12.7
FIXED
MOBILE except aeronautical mobile
BROADCASTING
BROADCASTING-SATELLITE

839 840 843 844 846</td>
</tr>
<tr>
<td rowspan="2">12.5 — 12.75
FIXED-SATELLITE (space-to-Earth) (Earth-to-space)

840 848 849 850</td>
<td rowspan="2">12.5 — 12.75
FIXED
FIXED-SATELLITE (space-to-Earth)
MOBILE except aeronautical mobile
BROADCASTING-SATELLITE 847

840</td>
</tr>
<tr>
<td>12.7 — 12.75
FIXED
FIXED-SATELLITE (Earth-to-space)
MOBILE except aeronautical mobile

840</td>
</tr>
</table>

Table 3.3 (cont'd)

836 In Region 2, in the band 11.7 — 12.1 GHz, transponders on space stations in the fixed-satellite service may be used additionally for transmissions in the broadcasting-satellite service, provided that such transmissions do not have a maximum e.i.r.p. greater than 53 dBW per television channel and do not cause greater interference or require more protection from interference than the coordinated fixed-satellite service frequency assignments. With respect to the space services, this band shall be used principally for the fixed-satellite service. The upper limit of this band shall be modified in accordance with the decisions of the 1983 regional administrative radio conference for Region 2 (see No. **841**).

837 *Different category of service:* in Canada, Mexico and the United States, the allocation of the band 11.7 — 12.2 GHz to the fixed service is on a secondary basis (see No. **424**).

838 In the band 11.7 — 12.5 GHz in Regions 1 and 3, the fixed, fixed-satellite, mobile, except aeronautical mobile, and broadcasting services, in accordance with their respective allocations, shall not cause harmful interference to broadcasting-satellite stations operating in accordance with the provisions of Appendix **30**.

839 The use of the band 11.7 — 12.7 GHz in Region 2 by the fixed-satellite and broadcasting-satellite services is limited to national and sub-regional systems and is subject to previous agreement between the administrations concerned and those having services, operating or planned to operate in accordance with the Table, which may be affected (see Articles **11**, **13** and **14** and Resolution 33).

840 For the use of the band 11.7 — 12.75 GHz in Regions 1, 2 and 3, see Resolutions **31**, **34**, **504**, **700** and **701**.

841 The 1983 regional administrative radio conference for Region 2 will divide the band 12.1 — 12.3 GHz into two sub-bands. It will allocate the lower sub-band to the fixed-satellite service and the upper sub-band to the broadcasting-satellite, broadcasting, mobile except aeronautical mobile, and fixed services, all services being on a primary basis.

842 *Additional allocation:* the bands 12.1 — 12.3 GHz in Brazil and Peru, and 12.2 — 12.3 GHz in the United States, are also allocated to the fixed service on a primary basis.

843 In the band 12.1 — 12.7 GHz, the Region 2 space services, existing or planned before the 1983 regional administrative radio conference for Region 2, shall not impose restrictions on the elaboration of the plan for the broadcasting-satellite service in Region 2 and shall be operated under the conditions set forth by that conference.

844 In Region 2, in the band 12.1 — 12.7 GHz, existing and future terrestrial radiocommunication services shall not cause harmful interference to the space services operating in accordance with the broadcasting-satellite plan to be prepared at the 1983 regional administrative radio conference for Region 2, and shall not impose restrictions on the elaboration of such a plan. The lower limit of this band shall be modified in accordance with the decisions of that conference for Region 2 (see No. **841**).

specifications. For instance, specifics regarding the inspection of stations and measures to be taken to eliminate interference between radio stations are put forth here. Part B of the Radio Regulations also contains a chapter on distress calls and safety communications. It is here in Part B of the Radio Regulations that the international signals used in case of distress — SOS and MAYDAY, pronounced as the French expression *"m'aider"* — are set forth. The Radio Regulations also contain more than 40 supplemental appendices. The official final document of an administrative radio conference is its Final Acts, which contain the Radio Regulations as amended and adopted by the conference.[39]

The Final Acts, like the Convention, can contain various resolutions, recommendations and reservations. The reservations are usually of a political, technical or formula nature. The formula reservation is the assertion by a country that it will take any steps "it deems necessary" to safeguard its interests in the event any other Member of the ITU fails to comply with the Radio Regulations.

3.5 CONCLUSION

The official creation of law and regulation of international space communication occurs in plenipotentiary and administrative radio conferences of the ITU. Such law and regulation is embodied in the International Telecommunication Convention and its annexed Radio Regulations.

Much of the law and regulation of international space communication, prior to its official adoption, evolves in preliminary form in official and unofficial settings associated with the preparation for and administration of the law-making conferences. The General Secretariat, the Administrative Council, and the Coordination Committee play a role in developing resources and establishing the context from which laws and regulations eventually arise.

Finally, two other permanent organs of the ITU also play a role, not only in preparation for law-making conferences, but also in implementation or interpretation of the laws and regulations that are adopted at those conferences. Those organs are the IFRB and the CCIR.

REFERENCES

1. "Current and Future State of Space Technology," *United Nations Document A/CONF.101/BP/2, March 16, 1981, p. 16.*
2. *International Telecommunication Convention, Nairobi, 1982,* Art. 4, (herein cited as *Nairobi Convention*).
3. *Ibid.*

4. *Ibid.*
5. *Ibid.*, Art. 5.
6. *Ibid.*, Art. 6; Plenipotentiaries of the ITU historically have been convened at eight to 10 year intervals — Montreux 1965, Málaga-Torremolinos 1973, Nairobi 1982; but because of the accelerating pace of change in conditions affecting global telecommunications, the Nairobi Conference modified Article 6 of the International Telecommunication Convention to shorten this duration. Even so, the next plenipotentiary planned is in 1989, seven years after Nairobi.
7. *Ibid.*, Art. 53.
8. *Ibid.*, Art. 16.
9. *Nairobi Convention, supra* note 2.
10. Wallenstein, Gerd D., "The I.T.U. Convention," in *International Telecommunication Agreements,* Dobbs Ferry, NY: Oceana Publications, 1979, p. 1.
11. *Nairobi Convention, supra* note 2.
12. *Ibid.*, Optional Additional Protocol.
13. *Ibid.*, Art. 6.
14. *Ibid.*, Art. 77.
15. *Ibid.*
16. *Ibid.*
17. Codding, George A. Jr., and Anthony M. Rutkowski, *The International Telecommunication Union in a Changing World,* Dedham, MA: Artech House, 1982, p. 66.
18. *Nairobi Convention, supra* note 2, Art. 77.
19. *Ibid.*
20. *Ibid.*, Arts. 56 & 77.
21. Codding and Rutkowski, *supra* note 17, at 162.
22. *Nairobi Convention, supra* note 2, Art. 56.
23. Codding and Rutkowski, *supra* note 17, at 172.
24. *Nairobi Convention,* Art. 8.
25. *Ibid.*, Art. 55.
26. *Ibid.*
27. *Ibid.*, Arts. 12 & 59.
28. *Ibid.*, Art. 59.
29. *Ibid.*, Art. 54.
30. *ITU Radio Regulations, 1982,* Art. 8.
31. *Nairobi Convention,* Art. 54.
32. *Ibid.*, Art. 7.
33. *Ibid.*, Art. 54.
34. *Ibid.*
35. *Ibid.*, Art. 10.
36. Codding and Rutkowski, *supra* note 17, at 69.

37. *Nairobi Convention,* Art. 1.
38. *ITU Radio Regulations, 1982,* Chaps. 3 & 4.
39. *ITU Radio Regulations, 1982.*

Chapter 4
Judicial and Advisory Organs of the International Telecommunication Union and Space Communication

4.1 INTRODUCTION

Fundamentally, the International Telecommunication Convention and Radio Regulations are self-enforcing. As noted in the Introduction and in Chapter 1, radio broadcast and reception equipment is essentially useless in the presence of harmful interference, particularly if such were to occur on a widespread or pervasive scale. The very nature of the radio spectrum and of radiocommunication technology dictates that cooperation and harmonization are in the vital self-interest of all concerned — a powerful incentive to promote compliance with international law still in the stage of development and lacking explicit enforcement mechanisms.

It is for this reason that the ITU has been so often pointed out as a particularly successful exercise in international cooperation. Yet, cooperation is not always available to solve every problem. Over the years some sort of at least quasi-judicial referee, a mechanism or organ for mediation and harmonization, became more and more necessary. The International Frequency Registration Board (IFRB) was formed to meet that need.

Even when there is a willingness to cooperate, cooperation may be frustrated by a lack of reliable information or by a lack of understanding of the nature of the resources and the technology involved. As radio technology evolved, the required cooperation and consultation became more and more complex. It thus became obvious that some sort of permanent advisory organ was necessary to engage in on-going technical consultations with Members of the ITU. The International Radio Consultative Committee (CCIR) is such an organ.

This chapter will examine the IFRB and the CCIR and their crucial functions in coordinating and harmonizing worldwide radiocommunication. The specific law and regulation of space communication in the ITU cannot properly be understood without at least a rudimentary understanding of how these organs contribute to the creation and enforce the execution of the specific provisions that relate to space communication.

4.2 THE INTERNATIONAL FREQUENCY REGISTRATION BOARD

The International Frequency Registration Board is the quasi-judicial organ of the ITU. The IFRB is charged with interpretation of the Radio Regulations and the application of those regulations to specific uses of radio frequencies that are notified to it with requests for recordation in the Master International Frequency Register.

The IFRB is described as *quasi*-judicial in that, like many other organs of international law, while it interprets the meaning and applicability of certain provisions of law, it has no power to enforce its findings. Yet, the impartial and judicious nature of the IFRB's findings are persuasive evidence of the international law embodied in the Convention and the Radio Regulations. Moreover, there is the added incentive of the moral force usually accorded instruments and organs of international law and procedure.

The respect accorded IFRB findings is heightened not only by the self-enforcing nature of the natural, technical requirement to cooperate in the use of radiocommunication, but also by the fact that sovereign nation-states usually prefer to resolve disputes without primary reliance upon binding arbitral or judicial processes. "Administrations are more interested in establishing interference-free communications than in perfecting legal claims to the occupancy of a particular frequency."[1]

Yet, because of the IFRB's quasi-judicial status, and because of the complexity, idiosyncracies, and inconsistencies of the Radio Regulations, the legal significance of the IFRB's functions and determinations is difficult to characterize or explain simply. It is perhaps best understood through its varied historical lineage, which has resulted in the Board becoming an unusual amalgam of several different functions, incorporating attributes of various types of institution.

In some sense the IFRB acts concurrently as a court, a recording office, a mediation and conciliation service, an administrative or regulatory agency, and a technical assistance agency. It resembles a court when it issues findings concerning the legal status of stations. It resembles a recording office when it enters into the Master International Frequency Register information concerning frequency assignments submitted by administrations. It resembles a mediation and conciliation service when it assists in the resolution of harmful interference disputes. It resembles a regulatory agency when it adopts interim technical standards and procedural rules. Finally, it resembles a technical assistance agency when it helps developing countries find frequencies for their new international communication requirements.[2]

4.2.1 Evolution of the Board

As noted in Chapter 2, the earliest provisions regarding notification of frequency assignments were established by the 1906 Berlin Radiotelegraph Conference to limit the possibility of interference between maritime stations. The 1906 Convention and its complement Radiotelegraph Regulations specified certain frequencies for use by maritime services and for services not available for public use, mainly military radio stations.[3]

The 1906 Regulations stipulated that maritime stations were to supply the Berne Bureau with a variety of information, such as call letters used, hours of operation, radio systems used, locations of coastal stations, and frequencies used.[4] The Bureau was to publish this information in the Berne List and to make it available to other administrations starting up new radio stations or changing already existing stations.

Beyond this simple requirement to furnish information, the 1927 Washington Radiotelegraph Conference established the basic approach to formulating the formal rules and obligations for regulating the use of the radio spectrum. The basic philosophy that was established has been reflected in the successive conventions and radio regulations since that time and can be characterized as an approach aimed at formulating "elastic provisions" with a minimum number of absolute rules and a maximum number of guides to action.[5]

Nevertheless, the 1927 Conference reaffirmed the general, abstract principle that stations of administrations must operate so as to avoid, so far as possible, harmful interference with other contracting governments. A table of frequency allocations, allocating all of the then usable frequencies to the different radio services, was adopted as a guide to action. Contracting governments were not obligated to adhere to the table. Still, the principle, if not the obligation, was to avoid harmful interference to stations of other countries. This principle qualified any broad claim of freedom to assign any frequency to any station.[6]

The 1927 provisions, substantially unchanged until 1947, provided a greater status to stations operating in allocated bands than to those operating "out of band." Yet, if no interference was caused, a station was not prohibited from using a frequency allocated to a different service, an approach still followed.[7]

The 1927 Conference also set modern precedent for the principle governing the use of the spectrum-orbit when it refused to recognize that any country had an absolute right to use a particular frequency. Yet, while joining in the formal refusal to acknowledge perpetual, priority rights to uses of radio, the United States submitted a proposal that was adopted that established in practice a kind of priority protection for first uses.

Section 16.1 of the 1927 Washington Radiotelegraph Regulations stated: "The frequencies assigned by Administrations to all new fixed, land or broadcasting stations which they have authorized or of which they have undertaken the installation must be chosen *so as to avoid so far as possible interference with international services carried on by existing stations the frequencies of which have already been notified to the International Bureau.*"[8]

4.2.2 Creation of the Board

As noted in Chapter 2, by 1947 there had been rapid technological advances in radio and a great increase in its use. Stations were operating on unregistered frequencies and others operated on frequencies that had been used by other countries before World War II. As a consequence of the chaos, the ITU looked into the possibility of replacing the Berne List with a new "engineered" frequency list, not based simply on a first-come, first-served preference, or an *a posteriori*, conditional priority.[9]

The concept of an engineered spectrum had its genesis long before 1947. The United States proposed a Universal Electrical Communications Union operating on the basis of sound engineering principles as early as the 1920 Washington Preliminary Conference.[10]

At the 1947 Atlantic City Administrative Radio Conference, the United States once again saw an opportunity to advance the system of radio spectrum management that had existed for more than 40 years. The United States proposed that all nations review their telecommunication needs and work with US experts to divide up the usable radio frequencies on the basis of sound engineering principles. The new plan would be maintained by a Central Frequency Registration Board, which could modify it as needs changed.[11]

The 1947 Conference scheduled various specialized conferences whose work it was to plan the use of the radio spectrum to the degree possible, according to the commonly agreed needs of the states of the world. A Provisional Frequency Board (PFB) was established by the Conference to draft a new International Frequency List (IFL) for the bands between 4000 kHz (4 MHz) and 27.5 MHz, engineered to the needs that could be identified at the scheduled conferences. The IFL would replace the Berne List and would serve as the repository of the plans of the engineered assignments.[12] As noted in Chapter 2, however, this *a priori* planning proposal eventually failed to be implemented because of inflated claims to spectrum usage and unsuccessful negotiations.

In 1947, the International Frequency Registration Board was established by the Atlantic City Administrative Radio Conference to help the PFB create the International Frequency List and also to coordinate ongoing international frequency management.[13] Although the work of the PFB to produce a fully engineered plan was not a great success, the joint effort with the IFRB did result in plans for the aeronautical and maritime services and for various radio services below 4 MHz.[14]

Yet, the ITU was still faced with the urgent need to revise the old list that had been maintained by the Berne Bureau to make it current with the Table of Frequency Allocations adopted at the Atlantic City Conference. Consequently, an Extraordinary Administrative Radio Conference (EARC) was called in 1951 to examine the situation in light of the failure of the PFB, and to implement the aforementioned plans developed for the aeronautical and maritime services and for various radio services below 4 MHz. EARC-51 also adopted plans developed by the PFB for the fixed services between 14 kHz and 150 kHz. Thus, by 1951, the first worldwide *a priori* plans were adopted, but they existed as mere fragments of the total allotment scheme that had been attempted. Consequently, the new International Frequency List was composed of frequency assignments whose arrangement cast but a shadow of the original vision of an engineered spectrum.[15]

4.2.3 Legal Foundations of the Board

The IFRB, as created in 1947, was charged with the legal and technical examination of notifications to it of frequency assignments and the issuance of findings with respect to the notifications. An **assignment** of a frequency to a radio station is made by individual countries. Frequency assignments are to be notified to the Board if the use of the frequency is capable of causing harmful interference to any service of another Member country, if the frequency is to be used for international radio communication, or if international recognition and protection is desired.[16]

Two new legal concepts were introduced into the Radio Regulations of 1947. These new legal principles referred to the *right to international protection from harmful interference* and the concept of *conformity with the Convention and Regulations*. It was the role of the IFRB to determine if notified uses of the spectrum were in conformity with the regulations. Only those stations in conformity with the Convention and Regulations were to be given international protection from harmful interference. This new aspect of the regulatory regime — conformity with the Convention and Regulations — stood as the formal recognition of criteria other than simply earlier notification to determine the rights and obligations of recorded assignments.[17]

Yet, while the IFRB was charged with overseeing the elaborate procedure for notification and registration of frequency assignments, it was not given any clear-cut legal distinctions to apply in the discharge of its duties. Only the legal status of one class of frequency assignment was clarified, and the new Frequency Register was designed to reflect this distinction.

In 1947, two new columns were added to the Frequency Register, the predecessor of today's Master International Frequency Register. The Registration column (2a) and the Notification column (2b) were added to reflect two classes of assignments: those in conformity with a frequency allotment plan in the Radio Regulations and those not in conformity.[18]

This reflected the basic assumption of the 1947 Conference that the radio spectrum would be engineered, resulting in international frequency allotment plans that would be adopted in succeeding years. An allotment plan is more specific than simply an allocation plan.

Allocations designate a group of frequencies to a service with priority *within* services decided largely on a first-come, first-served basis. **Allotments**, conversely, designate specific frequencies or bands of frequencies for the use of one or more identified countries or geographic areas under specified conditions, and therefore more closely resemble national assignments. An **assignment** is an authorization given by an administration for a radio station to use a particular frequency or frequency channel under specified conditions.[19] Indeed, the whole idea of the engineered spectrum was to correlate national assignments with international allotments.

An assignment in conformity with an engineered plan and with the Radio Regulations would be recorded in the Frequency Register with a date in the **Registration (2a)** column and would be entitled to international protection from harmful interference. Assignments operating in less than full conformity with the appropriate frequency allotment plan, but otherwise in conformity with the Regulations and Convention, would be recorded in the Register with a date in the **Notification (2b)** column, if it could be shown that there was no discernable harmful interference to already registered assignments in column 2a or 2b that had not caused harmful interference to others.[20]

Assignments that were unplanned or out-of-band and registered in column 2b would be given a lesser status than assignments with full international protection in column 2a. If a station caused harmful interference to a properly operating station, however, it was obligated to cease operations immediately, and would not even be registered by the IFRB unless its government made a demand for registration. Thus, unplanned operations were permitted only on a noninterference basis, with some degree of international recognition based upon the old first-come, first-served approach to frequency management preserved for stations that were not causing harmful interference.[21]

If a station were in conformity with the regulations and if it were in a planned band, but found to have a potential for causing harmful interference to already listed stations, the notification was to be returned to the administration. If the notification were resubmitted unmodified, it would be nonetheless registered in the notification (2b) column, with the date of original receipt listed. In essence, the Board had no power finally to refuse the registration of frequency assignments, all of which were anticipated in 1947 to be part of a frequency plan.[22]

The Atlantic City Conference also charged the IFRB with studying problems related to international frequency management if the Board was requested to do so by an administration. The Board was given the power to cancel frequency assignments that were not placed into operation within two years of notification. Notifications that had not been used for three years could be removed. These removals, however, could only be done with the agreement of the administration concerned. Hence, any fears that the Board might become too powerful were diminished.[23]

Implicit in the notification and registration process was the principle of the right to international protection from harmful interference. Yet, the practical meaning of this principle for protection of recorded assignments was not entirely clear.

Although the rights and obligations of recorded frequency assignments cannot be determined in the abstract, a general definition of protection has been given. "First, a recorded assignment is protected in the sense that the Board takes it into consideration when it examines a new notice for the probability of harmful interference to earlier recorded assignments. Second, a recorded assignment is protected in the sense that under certain conditions other assignments are obligated not to cause it harmful interference."[24]

Due, however, to an array of technical and legal distinctions potentially applicable to any recorded assignment because of a multitude of provisions in the Radio Regulations, as well as in reservations, resolutions, and recommendations of ITU radio conferences, any generalization concerning the rights and obligations of recorded assignments is hazardous. Nevertheless, although not explicitly stated in the Radio Regulations, the general concept of notification by date, coupled with the principle of international protection from harmful interference, seems to afford some protection, however limited and abstract, to every recorded assignment, at least as opposed to later registrants in the same class or registration column.[25]

In the event of disputes regarding harmful interference, the Board can be brought in to apply its settlement procedures if specifically requested to do so. The philosophy inherent in the Radio Regulations, however, is that administrations should make every possible effort to resolve their disputes on a bilateral basis and the IFRB should be brought in only as a last resort. Then the Board acts not as an arbiter but as a mediator, relying on a technical solution in which the legal status of the assignment, including the various dates listed in the Register, is not always the determining factor.[26]

Although the principle of international protection from harmful interference was laid down in the Radio Regulations, any dispute involving the rights and obligations that flowed from the provisions for notification and registration of a frequency assignment had to contend with an inherent and intentional vagueness of terms. In the end, resolution of differences required bilateral agreement between the countries involved.

Because of the original conception of a totally engineered spectrum, it was thought that most notifications of frequency assignments would be registered in the Registration (2a) column. Consequently, these notifications would enjoy the greatest level of international protection against harmful interference. The case, however, was and is otherwise.[27]

4.2.4 The Master International Frequency Register

By 1959 it had become obvious that the objective of establishing an engineered spectrum had to be abandoned. The 1959 General World Administrative Radio Conference, therefore, was faced with the task of adjusting a regulatory regime originally based on that objective.

As a consequence, the Conference adopted new provisions in Article 9 of the Radio Regulations that modified the IFRB's procedures to include most of the unplanned portions of the frequency spectrum. An elaborate procedure for notification and registration of frequency assignments was established in 1959 that required a redesign of the Frequency Register. Thus, the new Master International Frequency Register was created. The new Register would serve as the official central file for all recorded frequency assignments.[28]

The 1959 Regulations also formally established the Weekly Circular that had been authorized in 1947. This publication, which constitutes the official record of all frequency assignment notices received by the IFRB and its actions with respect to such notices, would be sent weekly to all administrations.[29] (See Table 4.1.)

In 1959, two classes of frequency assignments were created. While the registration 2a column and the notification 2b column were retained in the Master Register for assignments in the planned bands, a compromise provision that became part of Article 9 in the 1959 Radio Regulations also established a column 2d that was to be used for the unplanned bands. Henceforth, column 2d would be used to record the date on which a frequency assignment in the unplanned bands was received by the IFRB.[30] (See Table 4.2.)

The 1959 Radio Regulations, however, were unclear as to the legal status of column 2d assignments, especially with reference to the extent they enjoyed international protection from harmful interference. Although the concept of a fully engineered spectrum was abandoned in 1959, the legal distinctions upon which that objective was based were retained, even though they had become completely inadequate since most of the spectrum in use was unplanned.[31]

The changes to the Master Register came about as a result of the failure to make substantial alterations in the rights-vesting process through planning conferences where all states would define the available radio channels and geographic areas and equitably allot the radio resource among themselves. As a result, the overwhelming number of notifications would be listed in column 2d,[32] with whatever right to protection from harmful interference that existed relating only to subsequent registrations in the same column or category. Thus, rights were still secured primarily by an *a posteriori* vesting method that proceeded on a first-come, first-served, case-by-case basis, even though the highest degree of recognition and protection would continue to be obtained through planning conferences where the spectrum was planned on an *a priori* basis.

The Article 9 procedures adopted in 1959 for the notification and registration of frequency assignments have become even more complex, but, in substance, are the same as those in Article 12 of the 1982 Radio Regulations. These procedures, however, have been the basis of increasing contention within the international telecommunication community.

The basis of contention is that the practical effect of the consequent rights-vesting method is to give a *de facto* advantage to the administration

Table 4.1 Page from Weekly Circular of the IFRB
Source: International Telecommunication Union

UIT ITU

...ION INTERNATIONALE DES TÉLÉCOMMUNICATIONS | INTERNATIONAL TELECOMMUNICATION UNION | UNIÓN INTERNACIONAL DE TELECO...

COMITÉ INTERNATIONAL
D'ENREGISTREMENT DES FRÉQUENCES
IFRB

INTERNATIONAL FREQUENCY
REGISTRATION BOARD
IFRB

JUNTA INTERNACIONAL
DE REGISTRO DE FRECUENCIAS
IFRB

CIRCULAIRE HEBDOMADAIRE / DATE WEEKLY CIRCULAR / DATE CIRCULAR SEMANAL / FECHA	1776/02.06.87	SECTION SPÉCIALE Nº SPECIAL SECTION No. SECCIÓN ESPECIAL N.º	**ST61/1200**

ZONE EUROPÉNNE DE RADIODIFFUSION

Assignations de fréquence publiées aux termes de l'Article 4 de l'Accord régional pour la Zone européenne de radiodiffusion (Stockholm, 1961).

La presente Section Spéciale se compose des Parties indiquées ci-dessous par un × dans la case pertinente.

☐ *Partie A* - Modifications projetées aux Plans.

Partie B - Modifications insérées dans les Plans à la suite de ...oplication des dispositions de l'Article 4 de l'Accord.

...nulation d'assignations de fréquence dans les

...boles utilisés dans ... la Liste

EUROPEAN BROADCASTING AREA

Assignments published in application of Article 4 of the Regional Agreement for the European Broadcasting Area, Stockholm, 1961.

This Special Section is composed of the Parts indicated below by an × in the relevant box.

☐ *Part A* – Proposed modifications to the Plans.

☒ *Part B* – Modifications entered in the Plans as a result of th... application of the provisions of Article 4 of the Agreement.

☐ *Part C* – Cancell...ion of assignments in the Plans.

For the explanation of columns and symbols used in this Special Section, reference should be made to the Preface to the International Frequency ... (10th Edition – 1 February 1979) and to its Recapitulati...

ZONA EUROPEA DE RADIODIFUSIÓ...

Asignaciones de frecuencia publicadas en cumplimiento d... Acuerdo Regional para la Zona Europea de Radiodifusi... 1961).

La presente Sección Especial consta de las Partes indicada... una × en la casilla correspondiente.

☐ *Parte A* – Modificaciones propuestas de lo...

☒ *Parte B* – Modificaciones introducid... resultado de la aplicación de las dis... Acuerdo

☐ *Parte C* – Anulació... Planes.

Table 4.2 International Frequency List Showing Registration Columns
Source: ITU, *Radio Regulations*, App. 9, List I, Geneva, 1959.

	1		Assigned frequency (kc/s or Mc/s)
Date	2	2a	Of registration
		2b	Of notification
		2c	Of putting into use
		2d	Of receipt of the notice by the IFRB when columns 2a or 2b are not to be used
	3		Call sign (Identification)
Location of transmitting station	4	4a	Name of the transmitting station
		4b	Country in which the transmitting station is located
		4c	Geographical co-ordinates of the transmitter site (longitude and latitude) in degrees and minutes
Reception	5	5a	Locality(ies) or area(s) with which communication is established
		5b	Length of circuit (km)
	6		Class of station and nature of service
	7		Class of emission, necessary bandwidth and description of transmission
	8		Power (kW)
Transmitting antenna characteristics	9	9a	Azimuth of maximum radiation (ND if a transmitting antenna with non-directional characteristics is used)
		9b	Angular width of radiation main lobe
		9c	Antenna gain (db)
	10		Maximum hours of operation of the circuit to each locality or area (GMT)
	11		Megacycle order of the other frequencies normally utilized for the same circuit
	12a		Operating Administration or Company
	12b		Postal and telegraphic address of the administration responsible for the station
Remarks	13	13a	Results of examination and investigations by the IFRB
		13b	Remarks related to the finding by the IFRB
		13c	Other remarks

that first notifies a frequency assignment to the IFRB, with the latecomer having to negotiate from a position seeking accommodation. This contention is the basis for the argument of the developing countries that this type of *a posteriori* method of obtaining rights to use the GSO and frequencies associated with its use, works a particular inequity by perpetually freezing out those not yet ready to use such frequencies or orbital slots.[33]

It would not be until 1963, and further in 1971, that rights-vesting procedures for space radiocommunication were fashioned. Yet the basic notification-registration process, out of which all subsequent rights-vesting regulation for space radiocommunication has evolved, was put in place in 1959 and is still in place.

As can be seen, the IFRB was given few discretionary powers. It was empowered only to indicate in the list whether or not a notification was in accordance with the Convention and the Radio Regulations. Administrations, by virtue of the concept of national sovereignty, make the final decision regarding what frequencies they chose to use. If a country decides to proceed with the implementation of a frequency assignment that has been judged by the IFRB not to comply with either the Convention or the Regulations, the Board still has to record the notification in column 2d.

One reason Member countries might choose to ignore the findings of the Board is that the findings might be found not in their best interests. Since frequencies have been assigned primarily on a time priority basis rather than on "equitable principles," many states, especially in the early days of regulation, tried to hoard frequency assignments, some of which have been seldom if ever used. Consequently, difficulties still are encountered in the determination of frequency use.[34]

4.2.5 Structure and Function of the Board

The IFRB in 1987 was comprised of five independent members who are technically trained and possess a familiarity with a particular region of the world. The members are elected by the Plenipotentiary Conference from candidates sponsored by Member countries so as to "ensure equitable distribution among the regions of the world." The members serve not as representatives of their respective countries or regions "but as custodians of an international public trust." The effective dates on which the members begin their duties are determined at the time of their election and end at the decision of the following plenipotentiary.[35]

The members of the Board elect their own chairman and vice chairman who serve for one year. At the end of each year the vice chairman succeeds the chairman and a new vice chairman is elected.[36]

Prior to 1965 the IFRB consisted of 11 members. At the 1965 Montreux Plenipotentiary Conference an attempt was made to abolish the Board and to transfer its functions to a special unit of the General Secretariat. An influx of new, developing nations, however, who saw the IFRB as an ally against the developed nations in matters of frequency usage, defeated the proposal. Nonetheless, the Board was reduced to five members on the grounds that the 11-member Board was inefficient.[37]

The developing countries saw the Board as an important source of advice on problems of frequency management, as well as an independent, impartial protector of their interests, because it treated all frequency notifications in the same way. This position held by the developing countries has remained essentially the same.[38]

The 1982 ITU Convention charged the IFRB with these essential duties:

- to effect an orderly recording and registration of frequency assignments made by different countries as stipulated in the Radio Regulations with a view to ensuring "formal international recognition";
- to effect an orderly recording of the positions assigned by countries to geostationary satellites;
- to furnish advice to Members with a view to maximizing the number of radio channels and "with a view to the equitable, effective and economical use of the geostationary satellite orbit, taking into account the needs of Members requiring assistance, the specific needs of developing countries, as well as the special geographical situation of particular countries"; and
- to perform any additional duties related to the assignment and use of frequencies and the geostationary satellite orbit.[39]

The actual working arrangements of the IFRB are defined in the Radio Regulations. Essentially, the Board receives notifications from administrations regarding national frequency assignments, which the Board examines for conformity with the Convention and Radio Regulations. The regulatory procedures established for the space services, laid down in 1963, were comparable to those for the terrestrial services, although they were even more complex and entailed a coordination procedure that had to be undertaken before notification was made to the IFRB of a frequency assignment. The 1963 Extraordinary Administrative Radio Conference to Allocate Frequency Bands for Space Radiocommunication Purposes (Space Conference) introduced this coordination procedure.

Article 9A and Resolution 1A, adopted at the Space Conference, established a coordination procedure for earth and space stations that share with ***equal rights*** the same frequency bands with the fixed and mobile terrestrial services. Article 9A stipulated that a bilateral coordination procedure was required to be carried out prior to notification to the IFRB of an assignment to a fixed, mobile or space service sharing frequency bands on an equal basis.[40]

Resolution 1A of the 1963 Space Conference established the coordination procedure for international satellite systems, requiring an administration seeking to establish an international satellite system to provide the Board early in the system planning process with relevant information about the system. The Board then was to publish this data in its Weekly Circular for the information of all administrations. The procedure for coordination of international satellite systems was mandatory and had to be complied with before notification of a space station could be made to the Board under the Article 9A procedure.[41]

What was created is a three-step process: (1) advance publication; (2) coordination; and (3) notification. Information on planned satellite systems first is submitted for publication. After publication, coordination is undertaken. After coordination is completed, the IFRB is notified and the assignment is registered, or recorded, in the International Frequency Register.

The basic element of the 1963 regulatory framework for space radiocommunication consisted of a coordination procedure for the space services sharing frequency bands with terrestrial services, and it is still fundamental to the regulations for space communication. The process of bilateral coordination for the space services, however, is currently in question, and the possibility of replacing it with multilateral planning meetings was taken up at WARC-ORB-85. A more detailed discussion of that conference is found in Chapter 8.

4.3 THE INTERNATIONAL RADIO CONSULTATIVE COMMITTEE

The use of common telecommunication standards is a necessary prerequisite if governments and industry are to benefit in a complex, evolving telecommunication environment. The study of telecommunication innovations and the creation of common, international technical standards is the primary task of two permanent organs of the ITU, the International Radio Consultative Committee (CCIR) and the International Telegraph and Telephone Consultative Committee (CCITT). (The initials stand for their names in French, for example: CCIR — Comité consultatif international des radiocommunications.)

Matters dealing with space radiocommunication are handled by the CCIR, therefore our discussion concerning the consultative committees shall focus on this body. It should be kept in mind, however, that the purposes, structure, and duties of both consultative committees flow from the International Telecommunication Convention and are similar. In addition, occasionally there are matters that affect the work of both committees, and these matters are sometimes dealt with in Joint Study Groups.

4.3.1 Establishment of the Committee

As a result of the rapid advances in radio technology that had taken place by the end of World War I, the CCIR was created at the 1927 Washington Radiotelegraph Conference to study and to make recommendations on radio matters. Proposals for such a committee, however, had surfaced as early as the 1920 Washington Preliminary Radiotelegraph Conference. The CCIR was established to study technical and other questions related to international uses of radio and to present its conclusions to the next radio conference so that the conference itself would not be burdened with exhaustive, technical studies.[42]

It was not until 1947 at the Atlantic City Conference, however, that the consultative committees became permanent organs of the ITU. Since the committees were thereafter to meet regularly rather than periodically, provisions were adopted in 1947 that provided each committee with a permanent staff in the form of a specialized secretariat.[43]

4.3.2 Structure and Function of the Committee

The structure and function of the CCIR are enunciated in the International Telecommunication Convention. The duties of the CCIR are "to study technical and operating questions relating specifically to radiocommunication without limit of frequency range, and to issue recommendations on them. . . . " General economic issues are not to be addressed in these studies unless they involve comparing technical alternatives, and only then can economic factors be taken into consideration. Provision also is made for the study of questions and the formulation of recommendations "directly connected with the establishment, development, and improvement of telecommunication in developing countries in both the regional and international fields."[44]

The CCIR is open to participation by representatives of all Member administrations of the ITU. In addition, Recognized Private Operating Agencies (RPOAs) that have the approval of their administration and want to participate in the work of the committee may do so. This unique participatory blend of government and private enterprise has aided the consultative committees in gaining reputations for high-quality recommendations.[45]

The work of the CCIR is accomplished through its Plenary Assembly and its Study Groups, which are set up by the CCIR Plenary. The Plenary Assembly meets preferably every four years or at least eight months before a WARC that has been planned. The Director of the CCIR is elected by the Plenipotentiary Conference for the interval between two plenipotentiaries and is eligible for re-election at the following plenipotentiary. The specialized secretariat assists the Director. Laboratories and technical installations set up by the ITU also assist from time to time.[46]

The questions that the CCIR studies and subsequently makes recommendations on are referred to it by plenipotentiary conferences, administrative conferences, the Administrative Council, the IFRB, the CCITT, or by decisions made in the Plenary Assembly itself. In the interval between plenary assemblies, questions for study by the CCIR can also be requested or approved by correspondence of at least 20 Members of the Union. In addition, at the request of a country, the CCIR can also study and offer advice concerning its national telecommunication problems.[47]

Although the recommendations of the CCIR are not binding on administrations, the advice rendered by the committee is generally accepted and put into practice before it is adopted as part of the Radio Regulations. For this reason, recommendations of the CCIR on matters concerning radiocommunication are not taken lightly.

The Plenary Assembly decides on the questions to be studied and establishes the schedule for study groups. The material drafted in the study groups of the CCIR are reviewed at regularly scheduled interim meetings and the decisions are made whether these documents get forwarded to the Plenary Assembly as reports or as recommendations. If at any point in the decision-making process, an administration objects to the results reached, the study is resubmitted to the study group. At this point a representative from the objecting administration works with the group to clarify the purpose of the objection and to reach a workable solution. No recommendation ensues if an agreement cannot be reached, and the issue essentially is not considered further.[48]

It is in the Plenary Assembly that recommendations for international standards are made and resolutions are passed. Plenary assemblies also serve as the forum for establishing the technical data that become the basis for an administrative radio conference, such as the 1971 WARC for Space Telecommunications (WARC-ST).

4.3.3 Space Radiocommunication and the Committee

Under the impact of space communication, radio technical development tended toward a unified approach. Since the advent of space communication, many of the established specialized areas of radio have been challenged and essentially amalgamated. Until 1970, the CCIR operated with separate study groups that dealt with specialized branches of radio technology, such as transmitters, receivers, fixed service systems, and television. The 1970 Plenary Assembly approved a departure from this highly compartmentalized, traditional approach to allocating specific study tasks to groups on the basis of technical speciality, and adopted a service concept of study grouping.[49]

Under the service concept, a specialization is recognized to the extent that specific services may have specific requirements. Under the earlier concept, frequency range and the technology used more closely characterized a specialization. "Even so, the split of the Space Communication Study Group into one for scientific (SG-2) and another for commercial (SG-4) applications is regretted by many interested in a unified space technology treatment."[50] (See Table 4.3.)

What happened in 1970 was that the space astronomy and space research services were isolated from the more commercial services like television and telephony, even though the scientific and commercial services both depend on the same technology and frequency ranges. This isolation occurred when a separate study group for the space research and radioastronomy services was established.[51]

As the 1971 WARC-ST approached, the CCIR was pressed by demands for comprehensive, technical support. Thus, under pressure to act as a single-voice, radio-technical committee, the CCIR convened a Special Joint Meeting (SJM) of study groups. By this unprecedented move, all the groups interested in space radio techniques were brought together. Nine out of the eleven CCIR study groups participated in the work of the SJM. Just three months before the WARC-ST the SJM finished its work. The result was a 500-page report that had been prepared with the participation of experts from about 40 countries and from an even greater number of private operating agencies. This report served as the basis of the work and the principal source of reference for the 1971 Space Telecommunications Conference.[52]

Table 4.3 Terms of Reference CCIR Study Groups
Source: Wallenstein, Gerd D., *International Telecommunication Agreements*, Dobbs Ferry, NY: Oceana Publications, 1977. Reprinted with permission.

Study Group No.	*Before 1970*	*After the 1970 Plenary Assembly*
1	Transmitters	Spectrum utilization, monitoring
2	Receivers	Space research and radio astronomy services
3	Fixed service systems	Fixed service at frequencies below about 30 MHz
4	Space systems and radio astronomy	Fixed service using satellites
5	Propagation over the surface of the Earth and through the nonionized regions of the atmosphere	Propagation in nonionized media
6	Ionospheric propagation	Ionospheric propagation
7	Standard frequencies and time signals	Standard frequency and time-signal services
8	International monitoring	Mobile services
9	Radio-relay systems	Fixed service using radio-relay systems
10	Broadcasting	Broadcasting service (sound)
11	Television	Broadcasting service (television)
12	Tropical broadcasting	
13	Mobile services	
14	Vocabulary	(Vocabulary continued as an interim group, awaiting approval of a joint CCIR/CCITT Study Group)

4.3.4 Challenges Facing the Committee

Most of the participation in the work of the CCIR study groups is by administrations and private operating agencies of the developed nations. Different factors have been argued as the reasons for the absence of participation by the developing countries. One argument is the developing countries lack the finances to send delegates to the many study group meetings. Also, it has been suggested that the lesser developed countries (LDCs) cannot spare the trained personnel at home to follow the developments in the meetings of the consultative committees. Another reason given is that developing nations lack the expertise necessary to participate effectively at such a high technical level.[53]

The ITU, however, offers technical assistance, training, and other help to the developing countries to make it possible for them to be meaningful contributors. Commentators in the developed world have pointed out that many developing countries freely choose not to participate, even though a great deal of assistance is available to allow those who want to participate in the decision-making process to do so.[54]

Whatever the reason, the result is that, without the presence of many of the developing nations, the consultative committees have encountered problems in the satisfactory identification of questions and problems of specific concern to these countries. Another problem for the developing countries, particularly in the field of space communication, is that the consultative committees traditionally have been involved with the most advanced and expensive technologies. Often, the needs of the LDCs are not for these newer, more costly technologies, but rather for more affordable technologies that are older and less expensive.[55]

In any case, many developing countries tend not to participate actively in the work of the consultative committees. As a result, if the consultative committees are to continue to be the fora where standards for new and developing technologies are set, developed countries and developing countries must work together more closely to increase the participation of all concerned.

4.4 CONCLUSION

Both the CCIR and the IFRB are facing tremendous challenges. These two organs of the ITU were originally created to deal with matters concerning radiocommunication. As space communication developed, these organs of the ITU were charged with significant duties related to it. The great differences in technological development and culture between the developed and the developing world, however, challenged the original intent and definition of the organs.

For example, the underlying contention over the concept of equity between the developed and the developing world has also found its way into the debate over the procedures used by the IFRB. Where rapid technological advancement and the futility of attempts to engineer the spectrum have cooled the ardor of developed countries toward *a priori* planning, most of the developing world now clamors for it, especially in relation to space communication.

The geostationary satellite orbit particularly has been the focus of fears that the predominance of a first-come, first-served approach to frequency and orbital management will result in a perpetual loss of opportunity and access to space communication by the developing world. Over the next four chapters, therefore, efforts to allay these fears and to accommodate the need for more sensitive planning of the space services are recounted.

REFERENCES

1. Leive, David M., *International Telecommuncations and International Law: The Regulation of the Radio Spectrum,* Dobbs Ferry, NY: Oceana Publications, 1970, p. 25.
2. *Ibid.*, p. 24.
3. *Ibid.*, p. 41.
4. Codding, George A. Jr., and Anthony M. Rutkowski, *The International Telecommunication Union in a Changing World,* Dedham, MA: Artech House, 1982, p. 118.
5. Leive, *supra* note 1, at 44.
6. *Ibid.*, p. 45.
7. *Ibid.*, pp. 45-46.
8. *International Radiotelegraph Regulations, Washington, 1927,* 45 Stat. 2760 TIAS No. 767, Art. 5, Sec. 16(1), quoted in Leive, p. 47.
9. Codding and Rutkowski, *supra* note 4, at 272.
10. Leive, *supra* note 1, at 43.
11. Codding and Rutkowski, *supra* note 4, at 25.
12. Leive, *supra* note 1, at 64.
13. *Ibid.*, p. 57.
14. Codding and Rutkowski, *supra* note 4, at 33.
15. *Ibid.*
16. Leive, *supra* note 1, at 152. Whereas the requirement to notify the Board if a frequency assignment were capable of causing harmful interference to any other service or if the frequency were to be used

for international radio communication was established in 1947, the requirement to notify a frequency assignment to the IFRB to secure international recognition and protection was established in 1959.
17. *Ibid.*, pp. 56-57.
18. *Ibid.*, p. 60.
19. *ITU Radio Regulations, 1982,* Art. 1.
20. Leive, *supra* note 1, at 60.
21. *Ibid.*
22. *Ibid.*
23. Codding and Rutkowski, *supra* note 4, at 120.
24. Leive, *supra* note 1, at 82.
25. *Ibid.*, pp. 149, 160.
26. *Ibid.*, p. 129.
27. *Ibid.*, p. 60.
28. *Ibid.*, p. 326.
29. *Ibid.*, pp. 86, 329.
30. *Ibid.*, p. 70.
31. *Ibid.*, p. 71.
32. *Ibid.*
33. Codding and Rutkowski, *supra* note 4, at 277.
34. Smith, Delbert D., *International Telecommunication Control,* Leyden, Netherlands: A.W. Sijthoff, 1969, p. 30.
35. *International Telecommunication Convention, Nairobi, 1982,* Art. 10, (herein cited as *Nairobi Convention*).
36. *Ibid.*, Art. 57.
37. Leive, *supra* note 1, at 73.
38. *Ibid.*
39. *Nairobi Convention, supra* note 35, Art. 10.
40. *Final Acts of the Extraordinary Radio Conference to Allocate Frequency Bands for Space Radiocommunication Purposes, Geneva, 1963,* Art. 9A.
41. *Ibid.*, Resolution 1A.
42. Codding, George A. Jr., *The International Telecommunication Union,* Leiden, Netherlands: E.J. Brill, 1952, p. 121.
43. *Ibid.*, p. 460.
44. *Nairobi Convention, supra* note 35, Art. 11.
45. Codding and Rutkowski, *supra* note 4, at 105.
46. *Nairobi Convention, supra* note 35, Art. 58.
47. *Ibid.*
48. Codding and Rutkowski, *supra* note 4, at 99.
49. Wallenstein, Gerd D., "I.T.U. Case Studies," in *International Telecommunication Agreements,* Dobbs Ferry, NY: Oceana Publications,

1979, p. 7.

50. *Ibid.*
51. *Ibid.*
52. *Ibid.*, p. 8.
53. Codding and Rutkowski, *supra* note 4, at 104.
54. Correspondence to the authors from Stephen E. Doyle, Esq., chairman of the FCC Space WARC Advisory Committee for the ITU World Administrative Radio Conference on the Use of the Geostationary-Satellite Orbit and the Planning of the Space Services Utilizing It, May 25, 1987.
55. Codding and Rutkowski, *supra* note 4, at 105.

PART III
THE SPACE CONFERENCES

Chapter 5
ITU Conferences: 1959-1965 — Legal Framework for Space Communication Established

5.1 INTRODUCTION

The launch of the Soviet satellite Sputnik I on 4 October 1957 is generally heralded as the incident that marked the beginning of the Space Age. The launch of this 160-pound orbital vehicle, which circled the globe about every 96 minutes for a period of three months, also stands as the incident that sparked the development of international space communication law.[1]

As remarkable as Sputnik seemed to a world audience surprised and shocked by its launch, it is even more remarkable that the entire unmanned earth satellite program got under way in a peaceful and uncomplicated manner. What is extraordinary is that launches of earth satellites could have been subjected to serious delay due to international negotiation had many countries over which the satellites passed protested that their national sovereignty had been violated. Yet, the program was able to get under way without substantial protest.

Perhaps one of the most important elements enabling such a smooth inauguration of unmanned earth satellite programs, was that the principle of free flight for earth satellites had already been established internationally by the general consent of nations before Sputnik was ever launched. Those nations that did not give active encouragement to this concept tacitly consented by not voicing any protest.[2]

To understand how this principle of free flight for earth satellites became embedded into international law with such relative ease, we begin this chapter with a brief examination of historical events that led to its incorporation into international custom. These events relate to international cooperation in the study of our physical environment and once again demonstrate how change in scientific and technological capabilities drives change in the law.

5.1.1 Scientific Cooperation Advances International Law

Scientific and technological advance requires cooperation, both national and international. This is particularly true of studies and technical applications involving global phenomena. For example, the importance of obtaining geophysical data from remote areas of the world was recognized long ago. Yet, access to the various regions of the world that are controlled by many different governments is necessary for such activities to succeed.

More than a century ago, at the outset of the Industrial Revolution, and just before the discovery of radio waves in 1886, the first grand international attempt to study Earth on a large-scale basis occurred with the activities of the First International Polar Year. Through the cooperative activities of this extraterritorial endeavor that took place from 1882 to 1883, the first meteorological, magnetic, and auroral stations were established in the Arctic regions.[3]

It was proposed at that time that subsequent international scientific years be held every fifty years for the purpose of exploring the geophysical environment of Earth.[4] Consequently, 50 years after the First International Polar Year, a Second International Polar Year was held from 1932 to 1933. These two international endeavors contributed immensely to our understanding of the earth's magnetism and of the ionosphere.[5]

In 1950 a group of US scientists met and discussed the idea that since science had made such tremendous advances during and since World War II, another international year should be held before the lapse of the 50-year interval proposed at the time of the First International Polar Year.[6] This idea was eventually formulated into a proposal for an International Geophysical Year (IGY). The proposal was put forth by its directing body — the Special Committee for the International Geophysical Year (CSAGI). (The acronym CSAGI is from the initials of the name of the Committee in French.)

The CSAGI had a membership of representatives from various scientific unions including the International Union of Geodesy and Geophysics, the International Council of Scientific Unions, the International Astronomical Union, the International Scientific Radio Union, the World Meteorological Organization, the International Union of Geography, and the International Union of Pure and Applied Physics. The Joint Commission of the Ionosphere and the Joint Commission on Solar and Terrestrial

Relationships also were represented through these organizations. Thus, the International Geophysical Year, by virtue of its membership, was a larger undertaking than any of the previous cooperative international endeavors for the study of Earth's environment.[7]

In 1952 the CSAGI asked all countries to form national committees for the IGY and to submit recommendations for the program of the IGY. In 1953 proposals from 26 collaborating nations were considered and resulted in a tentative outline for the IGY program. National committees then began work on detailed planning of the activities of the parts of the IGY program for which they were responsible. In September and October of 1954, separate detailed programs that had already been prepared by about 38 nations were discussed in a CSAGI meeting so as to integrate them with one another. This meeting, held in Rome, produced a basic plan for the activities of the International Geophysical Year. The plan was adopted by all the delegations present.[8]

Ultimately 64 nations assisted in the IGY program as formal participants. Their contributions involved the wide use of both governmental and nongovernmental resources including the armies, navies, air forces, coast guards, government agencies concerned with standards, radar, radio, meteorology, coast and geodectic surveys, geological surveys, and a host of other scientific and research organizations. Universities and observatories in those countries where such institutions were not run by the State also were involved.[9]

The scientific mission of the IGY, the first new, transnational, cooperative venture since the founding of the United Nations, concerned areas of study characterized by their global nature: "solar activity, longitude and latitude, glaciology, oceanography, meteorology, geomagnetism, aurora and airglow, ionospheric physics, seismology and gravity, cosmic rays and upper atmosphere rocket studies, including the use of instrumented satellite vehicles."[10] By supporting a satellite program as part of the IGY, the participating countries agreed to the legal validity of the project and the necessary, associated principle of the free flight of earth satellites.[11]

It was agreed to hold the IGY from 1 July 1957 through 31 December 1958 in order to observe the entire period of maximum sunspot activity that was to occur during this time. World Data Centers were established around the globe, with primary data collection centers established in the US and the USSR and subcenters established in eight other countries. Tens of thousands of scientists and volunteers ultimately participated in the activities of the International Geophysical Year.[12]

Initially the CSAGI did not establish a satellite program. But the importance of upper atmosphere research was emphasized by one of the participating members — the International Scientific Radio Union (URSI). Subsequently, a Rocket Group was formed at the CSAGI meeting held in Rome from 30 September to 4 October 1954. A Rocket Program for the IGY was outlined that gave explicit support to the conduct of satellite experiments. Still other significant national and international scientific bodies also endorsed an earth satellite program.

On 29 July 1955 it was announced by President Eisenhower that the US contribution to the IGY would be just such a program. A satellite was to be launched under the auspices of the US National Committee for the IGY.[13] The satellite program was stimulated by the resolution passed by the CSAGI at its Rome meeting in 1954. It was planned that the developmental work would be completed in time for a satellite to be launched during the IGY that was planned for 1957 to 1958, a period of time when the international community would conduct worldwide observations in the fields of earth sciences.[14]

The US satellite announcement was well received by the world community, including the Soviets, who made a similar announcement within a week.[15] The announcement certainly was an historical event in that under existing international law a nation could have claimed that the start of such a program necessitated international agreement, without which the planning of a satellite launch over national territory might be suspect in terms of violating such territory.

In fact, the consent of so many nations to the IGY effort helped to establish in international law the principle of free flight for artificial earth satellites by authorizing the passage of satellites over subjacent territory.[16] Thus, the free passage of satellites over the territory of sovereign nations was by consent brought into customary international law. Now no nation can legally forbid such passage unless denial is based upon grounds of self-defense.[17]

5.1.2 Sputnik Challenges International Law

Although the technological strength exhibited by the Soviets with the launch of Sputnik I on 4 October 1957 was a shock to an unattentive world community anticipating a US satellite launch for the International Geophysical Year, its reception further strengthened the principle of free flight. The launch of Sputnik, however, while strengthening the free flight principle, concurrently exposed a major vacuum in international law regarding the use of radiocommunication in association with space activities.

Prior to the the launch of Sputnik, official definitions or regulations for space activities did not exist in either the Convention or the Radio Regulations of the ITU. Nor had any government officially introduced any definitions of the kind into international law. What activities were under way in this regard were under the aegis of the United Nations and generally related to international cooperation. Thus, as often is the case, the Sputnik launch highlighted the fact that technological development and utilization usually precede the regulatory environment. The use of space radiocommunication was no exception, as its first use actually occurred two years before any related national or international agreements were entered into.

It was not until the World Administrative Radio Conference of 1959 that specific provisions were made in the ITU Radio Regulations for radio allocations to the space and earth-space radio services. The 1959 WARC established the first international regulations for space radiocommunication, even though more than 73 radio transmitters were already operating on board 43 vehicles that were in orbit at that time, vehicles launched by the US and the Soviet Union.[18]

The focus of this chapter is the early response in the ITU to this new and unfolding development of space radiocommunication. The chapter covers the first plenipotentiary and administrative radio conferences that dealt with space radiocommunication during the early, experimental phase of its development. Although the 1959 conferences were confronted with many pressing problems other than space radiocommunication, our focus here is only on those regulations that relate to the space services, as we are primarily tracing the evolution of the current regulatory environment for space communication.

5.2 THE 1959 WORLD ADMINISTRATIVE RADIO CONFERENCE

When the states of the world came together for this first general radio conference with broad jurisdiction to review the entire Radio Regulations since the restructuring of the ITU in 1947, the struggles following the 1947 Atlantic City Conferences had to be recounted and reconciled.

The 1959 WARC was faced with the question of what to do with the objective enunciated in Atlantic City of creating a completely engineered frequency spectrum. So substantial was the preoccupation at WARC-59 with trying to develop a sufficient regulatory regime for radiocommunication that only minimal attention was given to developing a framework and allocating radio frequencies for the new, rapidly advancing needs of space radiocommunication. Nonetheless, the 1959 WARC did establish the first definitions for space radiocommunication services and allocated radio spectrum to them.

Radio plays an essential role in space operations and is paramount in the operation of space objects. It is by means of radio waves that the guidance, tracking, and monitoring of space vehicles is accomplished. Any two-way transmission of information between Earth and the spacecraft takes place by means of radiocommunication.

Richard Butler, Secretary General of the ITU, very succinctly stated this point before the Pontifical Scientific Academy at the Vatican. "Telecommunications provide the only link between space and earth and whatever happens in space or whatever use is made of space, telecommunications are required to make it possible." He added that if a legal framework were not observed by all nations of the world, reliable communications between space and earth would be impossible and "it would not be feasible to envisage projects and investments of the magnitude which we have already seen in the as yet rather brief period of man's exploration and use of space."[19]

5.2.1 Early Space Communication Regulation Established

Recall that Sputnik I was launched in October 1957. Many radio frequencies were used for transmissions to and from space for Sputnik and for all subsequent vehicles operating in space. Yet, until the Final Acts of WARC-59 were signed in December 1959, no radio frequencies were actually allocated by the Radio Regulations for space communication.

This did not mean, however, that transmissions to and from space during the interval from 1957 to 1959 were entirely unregulated. Use on a **noninterference basis** of frequencies that were unregistered or not in use was an entirely valid, legal, or "regulated" use. Use by satellites of frequencies that caused interference, however, was an unregulated use, and Sputnik did use a frequency allocated to another purpose and service — the international time frequency. In fact, there were a few scattered reports of interference from her signal.[20]

When WARC-59 adjourned on 21 December 1959, after nearly six weeks of meetings, the 1947 Radio Regulations had been revised to include new definitions for the space services, and frequencies were allocated for their use. Through revision of ***Article 1 of the Radio Regulations***, two new services were established by the WARC that met in Geneva in 1959: the **space service** and the **earth-space service**. Definitions of new terms such as space station and earth station also were adopted.[21]

With the revision of ***Article 5 of the Radio Regulations***, more than 125 MHz of the radio spectrum between 1,000 and 10,000 MHz were allocated in 13 frequency bands for space use. These allocations, however, were made for space research purposes only, with no frequencies allocated for such use as satellite communication. In total, about one percent of the radio spectrum was made available for space activities.[22] Nonetheless, the way was paved for the introduction of space radiocommunication services.

5.2.2 The 1959 Space Communication Provisions Inadequate

Although revising the 1947 Radio Regulations and overhauling the Table of Frequency Allocations were the basic tasks of WARC-59, the meeting also dealt with additional frequency demands. These demands were created by radio astronomy and requests for frequencies from organizations that needed communication channels for research purposes in connection with outer space.

Even though the ITU made significant strides in regulating activities in space, however, overall frequencies allocated to the space services in 1959 were inadequate. The allocations made were solely for the purposes of space research. Although the conference did establish the future of the work of the ITU in space radiocommunication, the allocations made were more in recognition of the existing uses of the radio spectrum than in anticipation of possible future requirements. The conference adopted measures required to solve only the most urgent problems. Thus, the needs of any planned or foreseeable astronautical programs were not met.[23]

This is not to say that the conference entirely ignored future needs. One of the notable accomplishments of WARC-59 was that the conference unanimously recommended that an Extraordinary Administrative Radio Conference be held in 1963 to examine the technical progress in the use of radiocommunication for space research and the results of technical studies by the CCIR and other bodies pertaining to space telecommunication.

The 1963 EARC, or Space Conference, was to decide which frequency bands to allocate to the various categories of space radiocommunication that were considered essential from the results of space research and other studies that had been conducted. Thus, although the accomplishments of the 1959 Conference were considered inadequate in terms of allocating sufficient frequency bands for space radiocommunication, the conference did establish the first definitions, allocations, and procedures for the establishment of space communication law.

5.3 THE 1959 GENEVA PLENIPOTENTIARY CONFERENCE

A plenipotentiary conference was taking place in Geneva in 1959 at the same time as the WARC was being held. The Geneva Plenipotentiary Conference, charged by the ITU Convention with the responsibility of revising the Union's constitution, adopted an important revision to the International Telecommunication Convention that would influence space radiocommunication. The 1959 Plenipotentiary Conference amended ***Article 4 of the ITU Convention*** to include a new function: to foster "the creation, development and improvement of telecommunication equipment and networks in developing countries by every means at its disposal. . . ."[24]

The developing countries insisted that this new provision be added to the constitution of the ITU. From 1959 forward, as the number of developing country Members increased, so did their role in the shaping of the Union. From that time forward also the role of the Union increased in facilitating the use of space radiocommunication in developing countries. The Article 4 amendment to the Convention served as formal acknowledgement of the responsibility of the ITU in the area of technical assistance activities. Although technical assistance activities were not constitutionally recognized until 1959, they had, however, been approved in 1951 by the Administrative Council.[25]

Article 4 paved the way for the developing countries to secure technical assistance from the ITU and its organs, especially from the IFRB. Pursuant to this 1959 provision, the developing countries have sought and obtained special assistance from the ITU in various aspects of telecommunication matters, especially in connection with space radiocommunication.

Article 13 of the ITU Convention also was amended in 1959. This revision required the consultative committees to "pay due attention to the study of questions and to the formulation of recommendations directly connected with the establishment, development, and improvement of telecommunication in new or developing countries in both the regional and international fields."[26]

The amendments of Articles 4 and 13 in 1959 were the first in a series of constitutional revisions over the years aimed at recognizing the special needs of and providing aid to the lesser developed countries. Consequently, a major activity of the ITU since 1959 has become development assistance. Prior to 1959, the ITU did not concern itself much with direct assistance to administrations to aid them in improving their telecommunication services. That task was left primarily to each individual state. Since the 1950s, however, the ITU has been increasingly involved in technical assistance to the lesser developed countries.[27]

5.4 THE 1963 EXTRAORDINARY ADMINISTRATIVE RADIO CONFERENCE TO ALLOCATE FREQUENCY BANDS FOR SPACE RADIOCOMMUNICATION PURPOSES

Since the mid-1960s the work of the ITU concerned with radiocommunication has been dominated by the rapid technological advances in space communication. Although the work related to space radiocommunication accomplished at the 1959 Conference was unquestionably limited, the conference did provide for the convening of the 1963 Extraordinary Administrative Radio Conference. The accomplishments of this 1963 "Space Conference" significantly advanced space radiocommunication by facilitating the introduction of commercial communication satellites and the establishment of global communication satellite services.

Yet, EARC-63 did not occur in a vacuum. As anticipated at WARC-59, there was tremendous space-related activity, both technical and political, between 1959 and 1963. The so-called "space race" between the US and the USSR had been joined in earnest, with both countries engaging in programs large and varied enough to put people into space.

5.4.1 Events Leading Up to the 1963 Space Conference

By 1961 the United Nations General Assembly was already considering matters related to space communication and acknowledged the ITU's responsibility for the province of space radiocommunication when it unanimously adopted Resolution 1721 (XVI) on International Cooperation in the Peaceful Uses of Outer Space. The resolution invited the ITU, in collaboration with other UN agencies, to consider requests from Member countries for technical and other assistance to make effective use of space radiocommunication, which should be available to all nations on a global and nondiscriminatory basis.[28]

These principles of providing for the use of space for international public telecommunication on a nondiscriminatory basis, as well as equitable access to space, were further reinforced by the passage in the US Congress of the Communications Satellite Act of 1962 (Comsat Act).[29] The significance of US involvement in establishing a global satellite communication network had been recognized by US Presidents Eisenhower and Kennedy.[30]

With the passage of the 1962 Comsat Act, US participation in the development and operation of an international satellite system was initiated, with US participation in the establishment of such a network to take place through a private corporation authorized by the Act — the Communications Satellite Corporation (COMSAT). Long before COMSAT could begin functioning as a business, however, certain assurances had to be made explicit in order that the planning of the future global satellite system and the large capital investment necessary for it could be made without fear of failure. Therefore, the US government began to concern itself with securing radio frequencies for space communication long before the 1963 EARC.

Although frequencies were allocated primarily for research purposes at the 1959 WARC, the United States did obtain tentative approval to use specific channels for experimental space communication. It was decided to consider the matter more fully in 1963. Consequently, at the 1963 EARC, the United States sought and secured approval for operational use of certain portions of the radio spectrum for space communication.[31]

From the viewpoints of the US and the USSR, frequency allocations for space communication were necessary for continued and progressive activity in the space environment. These nations also felt that frequency allocations for space communication should enjoy the same status as assignments to other services governed by the 1959 Radio Regulations. Specific allocations with permanent legal status were necessary, it was argued, if programs involving communications, meteorological, and navigational satellites, as well as space research, were to be assured of adequate frequency availability. It was the viewpoint of many developing countries, however, that such allocations should be regarded as interim and of lesser legal status until such time as a subsequent conference could establish "new principles indicating how the frequencies allocated to the space services could most equitably be apportioned among all administrations."[32]

Of course, the United States insisted that the frequencies designated for use by the space services be definite, sharing a status equal to other assignments. Eventually the US position prevailed, but as a result of the controversy, the conference was distracted from its primary task of reaching agreement on frequency allocations.[33] As a result of obtaining secure frequencies for use for space communication, COMSAT proceeded with its mission of establishing a global satellite network.

After incorporation, COMSAT, in collaboration with the US State Department and the Federal Communications Commission (FCC), began to stimulate interest within the international community in establishing a global satellite system. During 1963 and early 1964, negotiations were held, and in August 1964, the United States, COMSAT, and ten other countries entered into an agreement establishing an international consortium — the Interim International Telecommunication Satellite Organization — to be managed by COMSAT under the Interim Communication Satellite Committee (ICSC).[34] This interim organization evolved by 1973 into the permanent International Telecommunications Satellite Organization (INTELSAT).[35]

By 1987, the INTELSAT organization, which was permanently chartered in February 1973, had grown in membership to 112 countries. The prime objective of INTELSAT is to provide international public telecommunication via satellite to all areas of the world on a nondiscriminatory basis. Thus, the principle of nondiscriminatory, equitable access to the space segment for purposes of obtaining communication satellite service had been established in practice through the formation and operation of INTELSAT. Although it took ten years from the time of the 1963 EARC before INTELSAT was permanently chartered, the principles upon which the organization are based were laid in place at the UN in 1961, and were reinforced by the ITU at the 1963 Space Conference.

5.4.2 Legal Principles Enacted at the Space Conference

Four major items were on the agenda of the 1963 Conference:

1. to examine technical studies of the CCIR and other organizations in light of the technical progress made in space communication;
2. to allocate radio frequency bands considered essential by such studies for the various categories of space communication and for radio astronomy;
3. to revise the 1959 Radio Regulations as necessary for the successful implementation of conference decisions related to the allocation, notification, registration, and use of the space, earth-space, and radio astronomy services; and
4. to adopt additional provisions necessary to implement decisions regarding the technical characteristics of the space services.[36]

As the 1963 Conference began to grapple with its heavy agenda, it took into account several intervening factors such as the operational and flight experience gained in space since the launch of Sputnik I, and moves toward establishment of a global satellite system. Another element for consideration of the conference was the United Nations General Assembly Resolution 1721 (XVI) concerning freedom to use space and encouraging technical cooperation with developing countries. The resolution states that outer space is "free for exploration and use by all States in conformity with international law" and is "not subject to national appropriation." The resolution also encourages the ITU to give "sympathetic consideration to requests from Member States for technical and other assistance . . . so that they may make effective use of space communication."[37]

5.4.2.1 Full and Equal Legal Status for Space Services

As EARC-63 worked toward distinguishing and allocating suitable frequency bands for the space services, it also had to determine the procedure that was to be used for the notification and registration of these assignments. In this latter area, controversy was encountered. As has already been mentioned, the United States was promoting the development of a global satellite system through its legislatively authorized private corporation COMSAT. To assure success in this endeavor, the United States felt it absolutely necessary that the space services have the same definite and permanent legal status that assignments in other services enjoyed. The United States strongly opposed defining the legal status of assignments to the space services as interim, and thus of a lesser status than permanent assignments. The US argued that any frequency assignments to the space services should be given full legal status in order to establish a firm basis upon which to proceed with costly long-term programs. Ultimately, the US proposals were successful, however not without much effort.[38]

Opposition to the US position appeared in a proposal by Israel that reflected fears felt by the developing countries. Lesser developed countries were of the opinion that the frequencies available for space communication would be pre-empted on a national basis by the United States and other developed countries that would use their technological advantage to the detriment of the developing countries. The lesser developed countries feared that by virtue of the first-come, first-served principle inherent in the frequency notification and registration process, they would effectively be frozen out of the space segment. The Israeli proposal therefore sought to limit the frequency allocations for various space services by assigning them a provisional status, thus precluding any basis for claims of vested rights to such assignments. It was argued that it was the jurisdiction of a later conference to determine the precise legal status of any frequency allocations to various space services. The Israeli proposal, calling for the nondiscriminatory use of the frequency bands allocated for space communication purposes, advocated the abandonment or modification of the *a posteriori* rights vesting method of frequency management and claimed that the system of "individual frequency assignments to countries fails to ensure later the national interests of less developed countries."[39]

5.4.2.2 Equitable Access to Space Communication

Ultimately, the controversy was resolved when the conference adopted ***Recommendation 10A***, which recognized that all Union Members " . . . have an interest in and right to an equitable and rational use of frequency bands allocated for space communications." The recommendation further stated "that the utilization and exploitation of the frequency spectrum for space communication be subject to international agreements based on principles of **justice** and **equity** permitting the use and sharing of allocated frequency bands in the mutual interests of all nations."[40]

For the first time an international radio conference had recognized the principle of nondiscriminatory and equitable access to satellites for radiocommunication. The incorporation of this principle into international communication regulatory practice also indicated the beginning of a change in ITU policy orientation.

Although recommendations have no binding legal effect and are much more difficult to implement in practice than they are to draft, nonetheless, recommendations do indicate consensus within the ITU and offer some evidence of emerging trends. As will be discussed below, the essence of this recommendation enunciating the principle of nondiscriminatory, equitable access to space communication resources was later incorporated into the ITU Convention as revised in 1965 as Resolution No. 24. Thus, legal principles for space communication that had been discussed in global fora some years before the technology was widely accessible, formally became part of international communication regulatory practice just as the world entered the era of global space communication. (See Section 5.5 on the 1965 Montreux Plenipotentiary Conference for discussion of Resolution No. 24.)

5.4.3 Foundations of the Regulatory Regime for Space Communication

The 1963 EARC laid the foundation for the regulatory regime for space communication. During this conference, the 1959 Table of Frequency Allocations was revised to accommodate increased frequency allocations to the space services. The frequency bands allocated to these services jumped from one percent of the allocated spectrum in 1959 to fifteen percent in 1963, with allocations made on an exclusive or shared basis with the terrestrial services. Allocations were made in 51 frequency bands for the space services and totalled 6076.462 MHz. Of this total allocation for the space services, 2800 MHz of frequency bandwidth was allocated for communication satellites. Only 100 MHz of the frequencies allocated for communication satellites was on an exclusive basis, however, with the remaining 2700 MHz allocated below 10 GHz on a shared basis with terrestrial radio services.[41]

This was the first time that authorization for simultaneous use of frequencies in a common area by two different classes of services — space and terrestrial — was made. Although a major dispute did arise at the 1963 conference concerning proposals to define the space services in terms of their terrestrial counterparts, this approach was rejected.[42]

The sharing of frequencies on the basis of equal rights with stations in the terrestrial fixed and mobile services set a significant precedent for future radio resource management. It also necessitated the establishment of new procedures that were to be followed by the users of the shared frequency bands. These were referred to as coordination procedures and were incorporated into the ITU's Radio Regulations and linked with the IFRB, which was to oversee the coordination process.

5.4.3.1 New Space Services Defined

Before examining further the new regulatory regime established in 1963 for the space services, note that new definitions for various space services were also incorporated into the revised Radio Regulations of that year. ***Article 1*** was revised to include two categories of satellite communication — the communication-satellite service and the broadcasting-satellite service.

The communication-satellite service was defined as a space service: "between earth stations, when using active or passive satellites for the exchange of communications of the fixed or mobile service, or between an earth station and stations on active satellites for the exchange of communications of the mobile service, with a view to their retransmission to or from stations in the mobile service."[43]

Although the 1963 Conference also examined for the first time the regulatory problems raised by satellite broadcasting, the conference did not allocate any frequencies for its use. A definition, however, was incorporated into the 1963 revised Radio Regulations for such activities. The new space service was called the broadcasting-satellite service and was defined as "a space service in which signals transmitted or re-transmitted by space stations, or transmitted by reflection from objects in orbit around the Earth, are intended for direct reception by the general public."[44]

Thus, the accomplishments of the 1963 Space Conference were fundamental achievements necessary for the rapid introduction and use of commercial satellites for telecommunication. Without the legal definitions, frequency allocations, and sharing compromises established in 1963 for the space services, the institution of global commercial satellite communication might not have been so quickly achieved and the global village we know today could have taken longer to realize.

5.4.3.2 Coordination Procedure Established

Novel coordination procedures were of necessity introduced into the ITU's regulatory regime after the 1963 EARC decided that the space and terrestrial services would share frequency bands with equal rights. The rights-vesting procedures for space radiocommunication were initially fashioned in 1963. In general, the method adopted was similiar to the notification-registration procedure for frequency assignments to stations in the terrestrial services contained in Article 9 of the Radio Regulations. (The substance of the provisions relating to notification and registration of the terrestrial services and of the rights and obligations attached to them was outlined in Chapter 4 in discussing the IFRB's involvement.) Regulatory procedures for space services were similiar and of comparable complexity to previously enacted provisions for the terrestrial services.

A new procedure for the space services was introduced into the 1963 Radio Regulations, however. It was contained in ***Article 9A*** — "Notification and Recording in the Master International Frequency Register of Frequency Assignments to Stations in the Space and Radio Astronomy Services." The provisions incorporated into this article of the Radio Regulations were made necessary after the 1963 Space Conference concluded that space services and the terrestrial fixed and mobile services would share frequency bands between one and 10 GHz with equal rights. Technical and economical limitations made it mandatory to find allocations for the space services in this range. The new regulations required that a coordination procedure take place **before** an administration notified the IFRB or brought into service any frequency assignment to an earth station in a band allocated with equal rights to a space service and the fixed or the mobile service in the frequency spectrum between one and 10 GHz. It was required that an administration "shall effect co-ordination of the assignment with any other administration whose territory lies wholly or partly within co-ordination distance. . . ."[45]

Frequency assignments for the space services had to be notified to the Board under the same three conditions that applied to their terrestrial counterparts:

1. if the possibility for harmful interference existed;
2. if international communications were to be established; and
3. if international recognition of the frequency assignment was desired.

The method of calculating the coordination distance was set forth in Recommendation 1A of EARC-63. The coordination distance referred to the distance from a transmitting or receiving earth station within which the possibility of harmful interference existed between the station and another station in the fixed or the mobile service sharing frequency bands between one and 10 GHz.[46]

In general, the new regulatory framework for the space services involved publication by an administration of the intent to establish a station in a space service so as to elicit comments from other administrations whose own or planned stations might be subject to adverse interference from the proposed station or satellite network. A procedure of bilateral coordination for resolving any interference problem followed, to keep the estimated interference within limits prescribed in the Radio Regulations. In essence the notification-registration procedure, from which flows the right of international protection from harmful interference from subsequent assignments, thus extended the first-come, first-served rights vesting scheme, or priority principle, to the space services.[47]

The following discussion of the details of the coordination procedure illustrates the complexities involved in accommodating the space age and satellite communication. In addition to the technical complexities of accommodating satellite communication, there were also attendant political complexities.

The 1963 radio conference negotiations pertaining to regulating the space services were so complex that compromises had to be made to move any regulation forward. Thus, the requirements for prior coordination for earth stations were made different than those for satellite networks, with the regulations pertaining to satellite networks left intentionally vague.

No procedures or obligations comparable to those for earth stations, that is, coordination procedures for the proposed use of satellite systems with each other, with earth stations or with stations in the terrestrial services, were established in 1963. Proposals were submitted at the Space Conference that would have required formal coordination of a proposed space system with all other administrations, with respect to their existing or planned satellite systems. Such proposals, however, were opposed by the United States and other states who felt that they "would have permitted an administration to effectively block the space plans of another administration."[48]

The details of the coordination procedure can shed light upon the times and lay the groundwork for discussion of later radio conferences for space communication. Also, because a requirement for coordination, albeit one that differs slightly from that of 1963, still exists in the Radio Regulations, a brief examination of the procedure follows.

5.4.3.3 Coordination of Earth Stations

As revised in 1963, the Radio Regulations in Article 9A placed the strongest obligation to undertake the coordination procedure upon a country preparing to bring into use an earth station in the shared bands with the fixed and the mobile services. The coordination of such a proposed use was to take place prior to formal notification to the IFRB of the frequency assignment. Coordination consisted of various elements.

First the administration seeking coordination had to send a copy of a diagram identifying the location of the station and the proposed frequency assignment, along with all other related information required by the Radio Regulations, to any administration that was located within a specified distance — the coordination distance — and had earth stations that were in existence or planned. An indication of the approximate date when the proposed station would begin operations was also required.

Because the feasibility of space and terrestrial services sharing the same bands depended on the accuracy of the technical sharing criteria, such criteria as power limits, flux density limits for space station emissions, and minimum angles of elevation for earth stations were set out in Article 7 of the Radio Regulations and observance of specified criteria was made mandatory.[49]

According to Article 9A,[50] the IFRB was to be notified of the coordination process and was to inform all Member countries so as to provide an opportunity to any affected administration to be involved in the coordination process. The country with which coordination was sought was given 30 days to acknowledge the coordination data and to examine the information to see if harmful interference, as calculated by criteria established in the Radio Regulations, would result to any of its stations in the fixed or mobile services operating in conformity with the Regulations and Convention, or to such stations to be operated within the next two years. The administration was given 30 additional days to notify the administration requesting coordination of its agreement or disagreement.

In case of disagreement, a diagram detailing the location of relevant stations within the coordination distance, with all pertinent characteristics described, was to be sent to the administration seeking coordination along with suggestions aimed at solution of the disagreement. The IFRB also was to receive a copy.

The IFRB could be requested to assist in the coordination process if agreement could not be reached and under certain other circumstances: if a reply was not forthcoming within 90 days by the administration with which coordination was sought; if disagreement existed between the administrations as to the probability of harmful interference; and, finally, if coordination was impossible to effect between administrations for any other reason, such as the lack of diplomatic relations.

Even if the IFRB were unsuccessful in resolving any disagreement, however, the administration seeking coordination still could submit a notice to the Board requesting registration of the proposed operation, which the Board was required to register. The consequent legal status of the assignment, therefore, rested not on successful coordination but only on whether the obligation to coordinate had been satisfied.

If the 90 day period elapsed without a reply and the Board was brought in on the matter, the IFRB was to send a telegram to the administration with which coordination was sought. That administration then had 60 more days to reply. If no reply was received, it was to be considered that the administration had no complaint with respect to harmful interference that might be caused by the proposed earth station to any of its stations in the fixed and mobile services.

Thus, although the intention of the Article 9A regulations was to eliminate potential disputes in advance, no administration was obligated to agree to the proposed operation of another administration. There was only the obligation to make a good faith effort to do so. The consequences that resulted in the event an administration failed to reply to the IFRB within the 60-day period, however, did carry some legal weight, as well as practical consequences.

In essence, the administration with which coordination was sought waived whatever rights it possessed to subsequent complaint of harmful interference from the station of the administration seeking coordination. This right, however, was only lost if there was complete failure to respond. The right was not waived if some response was made, even if the response was viewed as inadequate or unacceptable by the Board or the coordinating administration.[51] Thus, the IFRB's lack of powers to resolve disputes between administrations was again evident.

Still, as made evident in more than three decades of practice, administrations gain no advantage in avoiding or declining coordination. It is in the interest of all interested administrations to coordinate to avoid harmful interference. This unavoidable reciprocity governs real life coordination and cooperation, almost regardless of the formula of the procedures. Two administrations cannot operate on the same frequency or on interfering frequencies without both administrations suffering a loss of the use, so they do not.

5.4.3.4 Coordination of Satellite Systems

As has been mentioned, no procedures or obligations to coordinate proposed satellite systems comparable to those for earth stations were established in 1963. Proposals to institute such procedures met with resistance. A compromise was adopted, however, in the form of ***Resolution 1A***. Application of the procedure established in Resolution 1A was restricted to "international satellite systems." The term, however, is nowhere defined. The proposal to restrict the application of the coordination procedure to international satellite systems was made by the USSR, "perhaps to ensure that the Resolution would not apply to its contemplated domestic system."[52]

Resolution 1A provided that an administration intending to establish an international satellite system notify the Board and submit relevant information as early as possible during the coordination process. The Board then was to publish this data in its weekly circular for the information of all administrations. Administrations that foresaw harmful interference to their existing or proposed satellite systems previously coordinated under

Resolution 1A were to comment within 90 days, with a view to reaching agreement. If no agreement was reached, the IFRB could be asked to intervene. If no response was ever forthcoming from an administration with which coordination was sought, it was assumed that there were no comments on the proposed action.[53]

The Resolution 1A coordination procedure was similiar to that for earth stations, however, unlike the provisions for the coordination of earth stations set forth in Article 9A of the Radio Regulations, the provisions of Resolution 1A were not binding upon Members and had no legal force. Also, other general weaknesses of the process have been cited. First of all, the resolution was restricted to international satellite systems. Consequently it did not pertain to domestic satellite systems. Additionally, there was no penalty established for complete failure to respond in the coordination procedure, as there was in the provisions for coordination of earth stations. An administration that failed to respond to a request to coordinate a satellite system did not waive its subsequent right to complain of harmful interference to its space services that might be caused by the proposed satellite system.[54]

Other weaknesses inherent in Resolution 1A were: it was intentionally imprecise on the data the administration seeking coordination had to provide regarding the proposed system; it lacked detailed provisions describing the obligations of administrations to coordinate and to respond, with the role of the IFRB in assisting in the coordination process vaguely described; it was unclear as to how disagreement between administrations was to be resolved; and it did not impose any obligation on an administration to accept any objections or to modify its proposed satellite system in any way.[55]

Although provisions for a coordination procedure to be carried out prior to notification to the IFRB of frequency assignments to space services sharing bands with the terrestrial services were first laid down in 1963, they were weak provisions which were the subject of revision in later years. At the next administrative radio conference for space communication, held in 1971, the conference specifically grappled with the notification-registration and coordination procedures. In cases involving the shared bands, the Radio Regulations were revised to include a procedure for the advance publication of information on planned satellite systems that was to take place prior to coordination of the proposed system. This element of the regulatory regime for satellite communication is elaborated in Chapter 6.

In 1963, the regulations for the space services were, if not more so, comparable in complexity to those put forth in Article 9 of the 1959 Radio Regulations for the notification-registration of the terrestrial services. Not only did the space and terrestrial services in the shared bands have to undertake the complicated notification-registration procedures to secure international protection from harmful interference, but also administrations were required to coordinate their proposed operations.

5.4.4 Preparation for the 1971 Space Telecommunications Conference

Although the foundation was laid in 1963 for international regulation of space communication, the provisions adopted raised many questions, some of which are just beginning to be examined in international radio conferences in the mid-1980s. Traditionally, international radio communication has been organized on a bilateral basis, which is solely dependent on the goodwill of the administration whose frequency is afforded international protection by virtue of prior notification and registration. Many communication satellite systems, however, are organized on a multilateral basis as satellite circuits can have a multipoint capability. Therefore, a major difficulty arose because the regulations governing satellite systems were based on the assumption of bilateral coordination. Consequently, administrations seeking to establish new national or international satellite networks had to rely then, and still must, on the goodwill of the administration responsible for a previously notified satellite network or networks in the same or potentially interfering bands for adjustments or accommodation.

No mandatory provisions or international agreements were established by EARC-63 requiring all satellite networks to make such adjustments and accommodations, beyond the practical reality that two administrations cannot use the same spectrum in the same geographical area simultaneously without coordination. Adjustments could encompass relocation of a satellite or modification of various system parameters, including reduction of power, modification of antennas and coverage areas, use of higher frequencies, and reduction of bandwidth.

The absence of mandatory provisions for accommodation can provide an advantage to existing services in the use of a particular frequency. The attendant technological and economical penalties associated with any adjustments were destined sometimes to be forced by established system owners to be carried by the newcomer.

It was from this position that the developing countries have held that they might be denied the benefit of equitable access to the space segment, even though the principle of equitability was enunciated in Recommendation 10A in 1963. Nonetheless, the 1963 regulatory approach, granting a certain right of priority to an earlier occupant of a particular frequency or orbital slot, over the years has become more and more a basis of contention between the developed and the lesser developed nations.

Thus, the stage was set at the 1963 Space Conference, not only for the future points of contention in regulation of satellite communication, but also for the subsequent evolution of those regulations. There was a realization at EARC-63 that responsive and on-going negotiations and conferences on satellite communication would be necessary. When the 1963 Conference adopted *Recommendation 9A*, it recommended that the Administrative Council annually review the progress of administrations in space radiocommunication and in light of this review recommend the convening of a radio conference to work out further agreement for the international regulation of space radiocommunication.[56] Thus, the 1971 World Administrative Radio Conference for Space Telecommunications had its genesis in 1963, and more work on space communication was scheduled for the near future.

5.5 THE 1965 MONTREUX PLENIPOTENTIARY CONFERENCE

5.5.1 General Provisions Amended

In 1965 ITU member countries met in Montreux, Switzerland, to consider revision of the ITU Convention. Two major aspects related to satellite communication were before the conference. One related to the previously enunciated legal principle of nondiscriminatory, equitable access to the space environment. The other related to technical and development assistance to the developing countries.

Between 1959 and 1965 the membership of the ITU had increased from 96 to 129. Most of this increase was a consequence of the influx of new, developing countries, which constituted a new majority and a drastic change in membership of the Union. A few of these countries were both very vocal and insistent on establishing principles and practices that would be of benefit to developing countries. [57]

In 1965 the Plenipotentiary Conference adopted ***Resolution 24*** entitled "Telecommunication and the Peaceful Uses of Outer Space."[58] This resolution reiterated the applicability of the principles set forth in the United Nations General Assembly Resolutions 1721 (XVI) and 1962 (XVIII). (See Chapter 9 for more elaboration on these resolutions.)

These UN resolutions had stated that telecommunication via satellite should be available to all states on a global and nondiscriminatory basis. In accord with the UN, the ITU considered "that, from the economic as well as the technical point of view, it is highly desirable that, for the full satisfaction of their needs, all countries should have equal opportunity to use space telecommunication facilities."[59] Thus, the Members of the ITU, by virtue of having adopted Resolution 24, were encouraged to promote the principle of nondiscriminatory and equitable access to the space environment for telecommunication purposes. This legal principle was eventually incorporated into Article 33 of the ITU Convention in 1973 and became a central theme in many proposals made by the developing countries related to space communication.

5.5.2 Resolutions, Recommendations, and Opinions

Another topic before the Plenipotentiary that had a bearing on space radiocommunication concerned technical and development assistance. The developing countries, unsatisfied with and critical of the assistance rendered heretofore in this area by both the ITU and the UN Development Program, came forth with what has been described as "three radical proposals."[60]

The developing countries proposed:

1. that an ITU development assistance fund be established separately from that of the United Nations;
2. that regional ITU offices be established in developing countries to aid administrations with their telecommunication problems; and
3. that a new organ of the ITU be created that would coexist with the CCIR and the CCITT and be concerned with matters related to technical assistance.[61]

The proposals were defeated. ***Resolution 28***, however, relating to development assistance activities, was passed at the 1965 Plenipotentiary Conference and was appended to the Convention. Although the developing countries had suffered defeat of most of their proposals, some of which would be incorporated into later conventions, a formal acknowledgement was made in 1965 of the importance of development assistance activities. Resolution 28 dealt with various forms of development assistance and technical cooperation. The resolution:

- invited technically advanced Member countries to facilitate visits by telecommunication personnel from the developing countries for instructive purposes;
- urged simplification of administrative procedures for approval of assistance from UN technical cooperation agencies;
- instructed the Secretary-General to assist in development assistance activities in various ways including collecting and disseminating information to the developing countries, exploring sources of financing for telecommunication plans in the developing countries, recommending to the Administrative Council the creation of posts that

could improve the services of the ITU to the developing countries, and annually reporting to the Council on the technical cooperation activities of the ITU;

- requested the Administrative Council to review all technical cooperation activities, to give more emphasis to holding and coordinating training seminars, and to review the organization of the Technical Cooperation Department; and
- recommended that the CCIR and CCITT consider ways of improving their operation to enable better response to questions from the developing countries.[62]

Thus, Resolution 28 adopted at the 1965 Plenipotentiary Conference was sweeping in nature and placed great emphasis on technical cooperation activities in the ITU. From this time forward technical assistance activities would be a major concern of the ITU.

Other resolutions were also passed at the Montreux Conference dealing with the ITU's role in development assistance. ***Resolution 29*** called for the recruiting of telecommunication specialists to help the developing countries with their telecommunication development and problems. It also directed the Secretariat to look into training methods for the personnel of the developing countries and to hold seminars for them and directed that the facilities of the ITU for providing developing countries with information and advice should be improved. All the permanent organs of the ITU were to aid in the dissemination of scientific knowledge and other advice to the developing countries. In addition, six other resolutions concerning technical cooperation were passed in 1965.[63]

Finally, ***Opinion 2*** was appended to the 1965 ITU Convention. Opinions set forth the consensus of attending delegates to a convention on certain issues. Opinion 2 was a statement "recognizing the advisability of organizing in the different regions of the world, through the United Nations Special Fund, centers for the study of space telecommunications . . . [and] that such centers should be established as soon as possible. . . ."[64]

5.6 CONCLUSION

Thus, the 1965 Plenipotentiary Conference, coming on the heels, as it did, of the first major space conference in 1963, naturally entailed much discussion related to space radiocommunication. The foundation for legal principles for space communication was strengthened at the 1965 Plenipotentiary Conference, and a major thrust toward technical and development assistance activities for the developing countries was launched. It would be over the next eight years, culminating at the next plenipotentiary conference in 1973, and following the Space Telecommunications Conference in 1971, that the principle of nondiscriminatory, equitable access to the space segment would be constitutionally recognized so as to create a binding, legal force upon all Members to adhere to the principle.

Some of the proposals made by the developing countries for development assistance would eventually also be passed in these later conferences. Thus, as the use of space radiocommunication grew, so did the role of the ITU in relation to it. As of 1965 that role was characterized by:

- Article 1 of the Radio Regulations, which included allocations for space services as early as 1959 and which in 1963 included allocations to the communication-satellite and broadcast-satellite services;
- Articles 4 and 13 of the 1959 International Telecommunication Convention on technical assistance for developing countries;
- Article 5 of the Radio Regulations, which established as early as 1959 allocations of frequencies for space services;
- Article 9A of the 1963 Radio Regulations on coordination of satellite systems;
- Resolution 1A of EARC-63 on coordination of international satellite systems;
- Recommendation 10A of EARC-63 on equitable access;
- Resolution 24 of the 1965 International Telecommunication Convention on equitable access;
- Six resolutions in the 1965 Convention on technical assistance for developing countries including Resolution 28 relating to various technical assistance activities and Resolution 29 relating to the training of personnel from the developing countries; and
- Opinion 2 of the 1965 Convention on regional centers for the study of space telecommunication.

These provisions provided the important framework of principles and practices upon which to build space communication regulation. Yet, they were still only a framework, a beginning, and the conferences of the 1970s and 1980s were bequeathed very challenging agendas.

REFERENCES

1. Haley, Andrew G., *Space Law and Government,* New York: Appleton-Century-Crofts, 1963, p. 128.
2. *Ibid.*, p. 67.
3. *Ibid.*, p. 63.
4. Robinson, George S., and Harold M. White Jr., *Envoys of Mankind: A Declaration of First Principles for the Governance of Space Societies,* Washington, DC: Smithsonian Institution Press, 1986, p. 40.

5. Haley, *supra* note 1, at 62.
6. Robinson and White, *supra* note 4, at 39.
7. Haley, *supra* note 1, at 63.
8. *Ibid.*
9. *Ibid.*, p. 64.
10. United States National Committee for the International Geophysical Year, *United States Program for the International Geophysical Year 1957-1958,* 1955, pp. 7-8, quoted in Haley, *supra* note 1, at 63.
11. Haley, *supra* note 1, at 64.
12. Robinson and White, *supra* note 4, at 40.
13. "Documents on International Aspects of the Exploration and Use of Outer Space, 1954-1962," *Staff Report, Senate Committee on Aeronautical and Space Sciences,* 88th Congress, 1st Session, Document No. 18, May 9, 1963, (herein cited as *Staff Report, Senate Committee, May 9, 1963*).
14. Haley, *supra* note 1, at 65.
15. Robinson and White, *supra* note 4, at 40.
16. Haley, *supra* note 1, at 68.
17. See generally, McDougal, Myres S., Harold D. Lasswell and Ivan A. Vlasic, *Law and Public Order in Space,* New Haven: Yale University Press, 1963; and Lay, S. Houston and Howard J. Taubenfeld, *The Law Relating to Activities of Man in Space,* Chicago: University of Chicago Press, 1970.
18. Haley, *supra* note 1, at 169.
19. Butler, Richard E., "Satellite Communications: Regulatory Framework and Applications for Development," *Space Communication and Broadcasting,* Vol. 3, 1985, pp. 103-108, at 103.
20. Correspondence to authors from Stephen E. Doyle, Esq., chairman of the FCC Space WARC Advisory Committee for the ITU World Administrative Radio Conference on the Use of the Geostationary-Satellite Orbit and Planning of the Space Services Utilizing It, June 16, 1987.
21. *ITU Radio Regulations, 1959,* Art. 1.
22. *Ibid.*, Art. 5.
23. Haley, *supra* note 1, at 174.
24. *International Telecommunication Convention, Geneva, 1959,* Art. 4.
25. Codding, George A. Jr., and Anthony M. Rutkowski, *The International Telecommunication Union in a Changing World,* Dedham, MA: Artech House, 1982, pp. 39-40.
26. *International Telecommunication Convention, Geneva, 1959,* Art. 13.
27. Codding and Rutkowski, *supra* note 25, at 283.
28. *United Nations General Assemby Resolution 1721 (XVI),* December 20, 1961.

29. Pub. L. No. 87-624, 76 Stat. 419, 1962, (codified at 47 U.S.C. Sections 701-704), *Comsat Act.*
30. *Staff Report, Senate Committee, May 9, 1963,* pp. 162, 208.
31. 1963 Space Conference, *U.S., Proposals for the Revision of the Radio Regulations, Document No. 8-E, June 12, 1963,* (herein cited as *U.S. Proposals, 1963*); *Final Acts of the Extraordinary Radio Conference to Allocate Frequency Bands for Space Radiocommunication Purposes, Geneva, 1963,* Art. 5, (herein cited as *Final Acts EARC-63*).
32. Leive, David M., *International Telecommunications and International Law: The Regulation of the Radio Spectrum,* Dobbs Ferry, NY: Oceana Publications, 1970, p. 209.
33. *U.S. Proposals 1963, supra* note 31.
34. *Agreement Establishing Interim Arrangements for a Global Commercial Satellite System,* (Intelsat Interim Agreement), signed *August 20, 1964,* 15 UST 1705, TIAS No. 5646.
35. *Agreement Relating to the International Telecommunications Satellite Organization "Intelsat"* (Intelsat Agreement), *August 20, 1971,* (entered into force *February 12, 1973*), 23 UST 3813, TIAS No. 7532.
36. Leive, *supra* note 32, at 209.
37. *United Nations General Assembly Resolution 1721 (XVI),* December 20, 1961.
38. Leive, *supra* note 32, at 209.
39. 1963 Space Conference, State of Israel, *Proposal to the Conference, Document No. 39-E, September 27, 1963,* (emphasis added) quoted in Leive, *supra* note 32, at 211.
40. *Final Acts EARC-63,* Recommendation 10A.
41. *Ibid.*, Art. 5.
42. Leive, *supra* note 32, at 210.
43. *Final Acts EARC-63,* Art. 1, Para. 84AG.
44. *Ibid.*, Para. 84AP.
45. *Ibid.*, Art. 9A, Para. 639AD.
46. *Ibid.*, Art. 9A, Para. 639AD.1.
47. Leive, *supra* note 32, at 243.
48. *1963 U.S. Delegation Report 27.*
49. *Final Acts EARC-63,* Art. 7.
50. *Ibid.*, Art. 9A.
51. Leive, *supra* note 32, at 224.
52. *Ibid.*, p. 225.
53. *Final Acts EARC-63,* Resolution 1A.
54. Leive, *supra* note 32, at 229.
55. *Ibid.*, p. 226.
56. *Final Acts EARC-63,* Recommendation 9A.
57. Codding and Rutkowski, *supra* note 25, at 44, 285.

58. *International Telecommunication Convention, Montreux, 1965*, Resolution 24.
59. *Ibid.*
60. Codding and Rutkowski, *supra* note 25, at 286.
61. *Ibid.*, pp. 54, 286.
62. *International Telecommunication Convention, Montreux, 1965*, Resolution 28.
63. *Ibid.*, Resolution 29.
64. *Ibid.*, Opinion 2.

Chapter 6

ITU Conferences: 1971-1977 — Legal Principles for Space Communication Extended

6.1 INTRODUCTION

The decade of the 1960s was a time when many technological and cultural changes took place in the United States and around the world. On an international level, the exploration of outer space and the use of space for purposes such as telecommunication were recognized to be the province of and for the benefit of all mankind. (See Chapter 9 for a discussion of the Outer Space Treaty.) Participation in the INTELSAT Consortium grew so substantially that by 1971, when the second world administrative radio conference was called to consider matters related to space communication, its membership included almost one-half of the ITU Members, giving such Members a direct interest in space telecommunication.

The substantial advances in space exploration and the associated technological developments and applications of space science that had been achieved by the United States, the Soviet Union, and other nations during the 1960s had the effect of changing the pattern and direction of human activity more than anything since electrical communication was introduced at the beginning of the industrial revolution. The world had entered the Space Age.

In particular, the successful launching of geostationary communication satellites and the development of stabilized, directional antennas for satellite use promised to revolutionize the entire field of communication. The application of space technology was already well known in other fields. The success of weather satellites had led to the implementation of a weather satellite system. Satellites began to be used for information gathering and as national technical means of verification of compliance with arms-control treaties and agreements. Space technology also offered the potential for development of a radio determination and traffic control system to meet aeronautical and maritime needs. What we now call the Information Age was beginning to explode.

So tremendous were the developments in space technology that the ITU Administrative Council in May of 1968 recommended that a WARC be convened during the latter part of 1970 or in early 1971 to consider the progress of administrations in space telecommunication. In *Resolution 632 of 1968*, the Administrative Council invited administrations to send proposals for the agenda of the conference and instructed the Secretary-General to report on the replies for consideration at the Council's next meeting in 1969.

Thus, the period between 1968 and 1969 was designated for the development of worldwide proposals for the agenda of the next radio conference concerned exclusively with space communication — the 1971 World Administrative Radio Conference for Space Telecommunications (WARC-ST). As a result of comments by ITU Members, the final agenda for the WARC-ST, which was quite broad in its mandate, was decided. The Administrative Council in May of 1969 then communicated this proposed agenda to the Members by circular telegram and received the necessary majority approval.[1]

6.2 THE 1971 WORLD ADMINISTRATIVE RADIO CONFERENCE FOR SPACE TELECOMMUNICATIONS

For six weeks during the summer of 1971 the nations of the world came together to decide how to accommodate the rapid advances that had been made in the use of geostationary satellites for radiocommunication. Recall that only modest provisions for space radiocommunication had been adopted by the 1963 Space Conference. As a consequence of the mandate of the 1971 Conference, the ITU was for the first time very visible on the world stage, responsible to facilitate achievement of the broad-range UN objective of placing space telecommunications within the province of and for the benefit of all mankind. Therefore, the task that was before the 740 delegates from 100 countries that met in Geneva in 1971[2] was substantial, eventually culminating in vast changes to the ITU technical standards and in its approach to the space environment.

The Final Acts of the World Administrative Radio Conference for Space Telecommunications were signed on 17 July 1971 and entered into force on 1 January 1973.[3] This formality, however, was merely the culmination of many years of concerted effort on behalf of many nations in the area of space exploration and of various organs of the ITU in space communications. The preparation for the conference alone entailed years of preliminary work.

After the Administrative Council first resolved to call a WARC for the consideration of space telecommunication matters, the CCIR began its preparatory work for the WARC-ST. The studies of the CCIR constituted the technical basis for the conference. The period between 1968 and 1971, however, was barely sufficient for adequate preparations.

Recall from Chapter 4 that in 1970 the CCIR set up an unprecedented Special Joint Meeting of Study Groups (SJM) to prepare for the 1971 WARC-ST, with a cross-referenced agenda that brought together all the groups interested in space radiocommunication. Nine out of 11 CCIR Study Groups worked together on matters related to space radiocommunication. Recall also that the creation and the convening of the SJM marked "the completion of an evolutionary cycle from specialization to unification."[4]

The report that emanated from the SJM contained the detailed technical findings and recommendations for the Space Telecommunications Conference. It contained the combined inputs of several hundred experts. Without such input the conference would have been hard pressed to accomplish its agenda — that of formulating international communication law to accommodate the developments that had taken place in space telecommunication.

Briefly, the 1971 Conference was charged with formulating regulations for satellite broadcasting and for other space communication services. The Table of Frequency Allocations was to be revised and supplemented as necessary. The conference was also charged with considering the feasibility of frequency planning for geostationary communication satellites and was to take any action that was deemed appropriate.

Perhaps the most controversial case before the WARC-ST was that of the broadcasting-satellite service (BSS), with the proceedings of the conference resulting in compromise aimed at deferring resolution of the matter until a subsequent conference. This chapter will examine in detail the issue of the BSS in 1971. The reader is referred to Chapter 9, however, for a comprehensive discussion of the overall negotiating history related to the BSS.

6.2.1 New Services Defined and Frequencies Allocated for Their Use

One of the significant accomplishments of the WARC-ST was the identification of many new services for telecommunication via satellite. Recall that in 1963 only two space communication services were defined for communication via satellite between points on earth — the communication-satellite service and the broadcasting-satellite service. Redefinition of some of the new space services in 1971 resulted in significant revision to ***Article 1 of the Radio Regulations*** and to the Table of Frequency Allocations in ***Article 5***, with consequent allocation of the radio spectrum increased to 275 GHz in 1971 from 40 GHz in 1963.

In the newly adopted service definitions of 1971, the *fixed-satellite service* (FSS) and the *mobile-satellite service* (MSS) replaced the communication-satellite service.[5] The redefinition of the FSS as adopted in 1971 was basically the same as that of today, the difference being that redundancies have been eliminated since 1971. Although the wording of the FSS definition was substantially changed from that of the communication-satellite service definition, it is not difficult to see how the one evolved into the other. (See Table 6.1.)

Table 6.1 Comparative Radio Regulation Definitions of 1963 and 1971 Showing Expansion of Fixed-Satellite and Broadcasting-Satellite Services*
Source: *Final Acts* of the 1963 EARC and 1971 WARC-ST.

1963		1971	
84AG	*Communication-Satellite Service* A space service: — between earth stations, when using active or passive satellites for the exchange of communications of the fixed or mobile service, or — between an earth station and stations on active satellites for the exchange of communications of the mobile service, with a view to their re-transmission to or from stations in the mobile service.	**84AG Spa2**	*Fixed-Satellite Service* A radiocommunication service: — between earth stations at specified fixed points when one or more satellites are used; in some cases this service includes satellite to satellite links, which may also be effected in the inter-satellite service; — for connection between one or more earth stations at specified fixed points and satellites used for a service other than the fixed-satellite service (for example, the mobile-satellite service, broadcasting-satellite service, etc.).

Table 6.1 (cont'd)

1963		1971	
84AP	*Broadcasting-Satellite Service* A space service in which signals transmitted or re-transmitted by space stations, or transmitted by reflection from objects in orbit around the Earth, are intended for direct reception by the general public.	**84AP Spa2**	*Broadcasting-Satellite Service* A radiocommunication service in which signals transmitted or retransmitted by space stations are intended for direct reception[1] by the general public.
		84AP.1 Spa2	[1]In the broadcasting-satellite service, the term "direct reception" shall encompass both individual reception and community reception.
		84APA Spa2	*Individual reception* (in the broadcasting-satellite service) The reception of emissions from a space station in the broadcasting-satellite service by simple domestic installations and in particular those possessing small antennae.
		84APB Spa2	*Community reception* (in the broadcasting-satellite service) The reception of emissions from a space station in the broadcasting-satellite service by receiving equipment, which in some cases may be complex and have antennae larger than those used for individual reception, and intended for use: — by a group of the general public at one location; or — through a distribution system covering a limited area.

*In 1971, the Fixed-Satellite Service replaced the Communication-Satellite Service.

The 1971 Conference also further distinguished between the newly defined fixed-satellite service and the broadcasting-satellite service by revising the definition of the BSS and allocating for the first time frequencies for its use. Specifically, the definition of the broadcasting-satellite service was slightly shortened from that of 1963 to read: "A radiocommunication service in which signals transmitted or retransmitted by space stations are intended for direct reception by the general public."[6] (See Table 6.1.) A footnote to the definition stipulated that "(i)n the broadcasting-satellite service, the term 'direct reception' shall encompass both individual reception and community reception."[7]

Individual reception was defined as: "The reception of emissions from a space station in the broadcasting-satellite service by simple domestic installations and in particular those possessing small antennae."[8]

Community reception was defined as: "The reception of emissions from a space station in the broadcasting-satellite service by receiving equipment, which in some cases may be complex and have antennae larger than those used for individual reception, and intended for use:

—by a group of the general public at one location; or
—through a distribution system covering a limited area."[9]

The feature distinguishing the BSS from the FSS as defined in 1971 was the flow of signals through the satellite from either one point to another point (FSS) or from one point to an indeterminate multitude of points (BSS). Included under the broadcasting-satellite service concept are direct broadcasts of sound or television to small, individual receivers and satellite transmissions of information to larger community reception points, sometimes for redistribution to individual receivers. BSS signals are intended for distribution to serial points simultaneously. In practice, of course, the distinction is not always so clear. The FSS has been the most widely used service, and has carried transmissions from one point to a multitude of other "fixed" points in a pattern strongly resembling broadcasting. Perhaps the real distinction between the FSS and the BSS is the different type of technology used, different levels of power, and the different range of frequencies allocated for each respective service.

In total, the 1971 WARC-ST allocated 97 frequency bands for the use of various space services,[10] making available about 15,000 MHz of new spectrum for the space services. These included bands for broadcasting, maritime, mobile, earth exploration, and other satellite services.

Frequency allocations to the BSS, which were the province of Committee 5 on Frequency Allocations, were the subject of extensive debate. Much of the debate focused on the fear of potential interference from the BSS because various administrations made proposals requesting that frequencies for the BSS be allocated in the lower frequency bands. Basically, these proposals were made on the basis of economics, as it was more economical to use the latest technology that operated in the lower frequency bands than it was to devise new systems. Also, the lower bands were considered preferable because of lower transmission losses and better overall propagation.

In the end, six frequency bands were allocated for use by the BSS, with only the 620-790 MHz and 2500-2690 MHz frequency bands allocated in the lower regions of the spectrum. The other allocations to the BSS were as follows: 11.7-12.2 GHz in Regions 2 and 3, 11.7-12.5 GHz in Region 1; 22.5-23 GHz; 41-43 GHz and 84-86 GHz. The allocations in the 22.5-23 GHz band were for Region 3 only, subject to agreed power flux density limits for the protection of the terrestrial services. The only exclusive, worldwide allocations made for the BSS were in the 41-43 and 84-86 GHz bands — bands that were not in practical use. The rest were allocated on a worldwide, primary shared basis.[11]

A service that was initially defined in 1971 was the earth exploration-satellite service. Because of the potential benefits that could be achieved from this new service, such as better earth-resource management in the fields of geology, geodesy, hydrology, forestry, and agriculture, the conference made worldwide allocations in the most sought after bands of the ultra high frequency (UHF) (300 to 3,000 MHz) and super high frequency (SHF) (3 to 30 GHz) portions of the spectrum for its use. The allocations made in these bands were made on a secondary shared basis with other services. They included the 1525-1535 and 8025-8400 MHz bands, with the 21.2-22 GHz band allocated to the earth-exploration satellite service on a primary basis with the fixed and mobile services. Today these bands are shared by microwave relay systems, UHF television, weather satellites, meteorological aids, radio astronomy, amateur radio, and other services. Other UHF bands were also designated for use by the earth exploration-satellite service on condition that no interference be caused to other stations operating in accordance with the frequency allocations table. These allocations were made in the 401-403, 460-470, and 1690-1700 MHz bands. A few allocations were also made to the earth exploration-satellite service in the SHF and extremely high frequency (EHF) (30 to 300 GHz) bands on a primary, shared basis with space research.[12]

Besides incorporating into the Radio Regulations new definitions and allocations in 1971 for the space services, frequency allocations for other services were also revised. These included meteorology, space research, the amateur service, the standard frequency and time signal services, radio astronomy, and the aeronautical mobile service. Also, the maritime mobile-satellite service was recognized and frequencies were allocated for its use.[13]

The frequencies allocated to the maritime mobile-satellite service were fundamental to the establishment of the International Maritime Satellite Organization (INMARSAT). INMARSAT is an international organization, established in 1976 on the model of INTELSAT, to provide for, administer, and manage a worldwide maritime satellite system that offers space radiocommunication services for radio navigation and for high quality, reliable communication.

The maritime mobile-satellite service was defined as "a mobile-satellite service in which mobile earth stations are located on board ships." Survival craft stations and emergency position-finding radiobeacon stations were also able to participate in this service.[14]

As is apparent, the work of the 1971 Conference in defining new space services and in allocating frequencies for their use was substantial and contributed to the development of all such services. Henceforth, space telecommunication in all its varied applications would be internationally recognized and would begin to expand into significant global applications.

6.2.2 Revision of Coordination, Notification and Registration Procedures

Issues having to do with the acquisition for orbital use of radio resources, now referred to as spectrum-orbit use, were among the major concerns at the WARC-ST. Some even framed the debate over fair use of and access to the spectrum-orbit in terms of "rights." Yet, the ITU actually grants or confers no rights. Uses are registered and legal recognition and international protection from harmful interference may be granted, but no right in any property sense is granted. Yet, legal recognition is often thought of and expressed in terms of rights, which has been true in the debate over regulation of the spectrum-orbit.

A principal focus of the 1971 Conference was the revision of ***Article 9A of the Radio Regulations*** governing the coordination, notification and recording in the Master Register of assignments in the space and radio astronomy services, except the broadcasting-satellite service. Results of the work in this area established the international framework for coordinated use of the geostationary arc. The 1971 Conference amended the Radio Regulations to restrict unilateral action in space telecommunication. This was effected by the adoption of a mandatory procedure for advance publication of information on planned satellite systems that was to take place prior to coordination.

Committee 6 of the conference, the Regulations Committee, worked closely with the Allocations and Technical Committees to establish a legal regime for space telecommunication. The Technical Committee was responsible for revisions of the Radio Regulations pertaining to frequency sharing between space and terrestrial services and for establishment of technical criteria for frequency sharing between space systems.[15] The points of the agenda were distributed over these three committees, with each having a significant part to play in the work of revising the Regulations to accommodate the burgeoning use of space for telecommunication.

In light of the progress that had been made in this area in the intervening years between space radio conferences and the consequent increase of use of geostationary satellites, revision of Article 9A became necessary. To ensure equitable, effective, and economical use of the radio spectrum and the GSO, a mandatory "Procedure for the Advance Publication of Information on Planned Satellite Systems" was adopted. It provided that a Member administration ". . . which intends to establish a satellite system shall, prior to the co-ordination procedure . . . , send to the International Frequency Registration Board not earlier than five years before the date of bringing into service each satellite network of the planned system, the information listed in Appendix 1B."[16] Today this provision reads: ". . . not earlier than five years and preferably not later than two years before the date of bringing into service each satellite network of the planned system, the information listed in Appendix 4."[17]

Appendix 1B, entitled "Advance Publication Information to be Furnished for a Satellite Network," set forth the basic operational data that was to be provided to the IFRB on any planned satellite system. The data required on orbital information was essentially the same as that required today. Members were obliged to furnish the IFRB with orbital information relating to space stations if the space station was situated on board a geostationary satellite. This was to include the "planned nominal geographical longitude on the geostationary satellite orbit and the planned longitudinal and inclination tolerances." Also to be indicated was the arc of the geostationary satellite orbit "over which the space station is visible, at a minimun angle of elevation of 10 degrees at the Earth's surface, from its associated earth stations or service areas. . . ."[18] Other operational details, such as frequency range, power characteristics of transmission, and antennas, were also required to be furnished in the advance publication phase.

If the calculation of risk of interference, using those characteristics of the system given during the advance publication phase, showed a potential for interference exceeding two percent of the total noise in any particular link, then coordination was required to take place with the affected administrations. The exact amount of interference was calculated during the coordination procedure, which had to be effected prior to final notification and not more than three years before the date of implementation of the system.[19]

The process as set forth in the revised Article 9A of the Radio Regulations was intended to be flexible, allowing for voluntary relocation of existing space stations to accommodate new space stations if such new stations could not be otherwise accommodated.[20] Thus, the revised provisions represented a step forward in the sharper demarcation of diverse forms of system arrangements and applications and placed heavy emphasis on international cooperation and accommodation, even to the extent that, if required, a relocation of a prior, existing satellite might voluntarily take place.

Recall that the coordination procedure as devised in 1963 was required for space services sharing frequency bands with equal rights with the terrestrial services between one and 10 GHz. This procedure, carried over to 1971 but made applicable to all shared frequencies above one GHz, was predicated upon the concept of sharing. Matters related to this concept accounted for much of the work of Committee 4, the Technical Committee. Committee 4 received these terms of reference: "To consider, revise and supplement, as necessary, the existing provisions of the Radio Regulations pertaining to the technical criteria applicable for frequency sharing between space and terrestrial services, and to establish technical criteria for frequency sharing between space systems. . . ."[21]

Sharing of frequency bands by different services enables efficient use of the frequency spectrum. For many services, however, the use of the lower frequencies is more attractive because the propagation conditions are more favorable. As the evolution of radio has been that the lower frequencies have been available for use first, older more established services dominate these lower frequencies. At the time of the 1971 WARC-ST, the desirable lower frequencies applied to the spectrum range from one to 10 GHz. For the BSS this was still lower and included the 80 MHz and up range. Therefore, numerous proposals for sharing were put forth at the 1971 Conference, necessitating the formulation of sharing criteria that would be accommodating to both old and new users of the same bands.[22]

Various issues are involved in sharing. The transmission power limits and the positions of stations relative to each other are technical issues that must be considered. By specifying the electrical characteristics and physical locations of stations, frequency sharing can be evaluated and planned. Different services, however, are constrained in different ways in their freedom to specify the various characteristics and locations. For instance, the radio astronomy service relies on very sensitive and selective equipment to receive cosmic signals of very low levels. This requires a special level of protection for frequency bands allocated to this service. Similarly, certain radio frequencies are absorbed by the atmosphere or are subject to substantial distortion; consequently, they are of limited or no use in remote sensing satellite systems. Still, other frequencies are readily absorbed by water molecules in the atmosphere and are of limited value in tropical rain zones.

The critical problems with which Committee 4 of the WARC-ST had to deal concerned sharing of frequencies by terrestrial and space systems and among space systems. Power limits imposed on the transmitters had to be chosen so that interference, or unwanted signals, would not exceed an acceptable level at the receiver site of the sharing service. These power limits, which refer not only to earth stations but also to the power flux-density produced by a space station's transmitted signal at the earth's surface, were only one problem that Committee 4 had to consider. Other areas of concern involved: definitions, angular limitations for stations of the terrestrial services and earth stations, the coordination distance procedure, sharing between space systems and other various assignments.[23] The work of the Technical Committee was fundamental to the success of the Regulations Committee in establishing regulatory provisions for space telecommunication — provisions that were based upon the sharing criteria formulated in the Technical Committee.

With reference to sharing between geostationary space systems, the additional problem of orbital slots had to be considered as only a limited number of satellites in the GSO can operate on the same frequency. Antenna sidelobes can create interference — areas of stray radiation in the vicinity of the antenna. This is an example of the type of spectrum usage consideration that has to be taken into account in orbital coordination and management. Sidelobe radiation presents a greater problem for satellite stations than for earth stations. It is highly unlikely that earth station antenna sidelobes will affect other earth stations, because transmitting earth stations are rarely close enough together, or may be separated by shielding or terrain features. A transmitting earth station with unconstrained sidelobe radiation operating in a shared band with terrestrial services, however, could cause interference in the terrestrial service in the area within a few miles of the earth station.

Antenna sidelobe radiation on satellites is even more critical and must be kept as small or as tight as possible by good engineering. Satellite proximity in space can be significantly influenced by the antenna sidelobe radiation of other satellites. Satellite spacing also can be affected by the size of transmitting antennas on the ground, as well as the transmitted signal strength, and the width of the earth station beam. Recall that the larger the antenna's diameter, the narrower the transmission beam can be, thus facilitating more orbital parking spaces. This also can be facilitated by the use of spot beams by satellites. WARC-ST, however, only acknowledged by reference the various technological and economic trade-offs that must be made for efficient and effective spectrum-orbit use and deferred the coordinated international planning and regulation of the GSO to the work of future conferences.

Although specific regulations for the GSO were not adopted in 1971, the conference approached the regulation of the geostationary orbit by adopting the traditional Article 9A notice and recordation procedure for use for satellite networks. This did not sit well with many developing countries. To allay fears that developing countries might be frozen out of the space segment because they were not yet technologically nor economically able to use it, the conference adopted Resolution No. Spa 2-1.

6.2.3 Resolution Spa 2-1 and Recommendation Spa 2-1

With the adoption of ***Resolution Spa 2-1***, entitled "Relating to the Use by All Countries, with Equal Rights, of Frequency Bands for Space Radiocommunication Services," Members of the ITU were called upon to take into account "that the radio frequency spectrum and the geostationary satellite orbit are limited natural resources and should be most effectively and economically used." It was noted that "all countries have equal rights in the use of both the radio frequencies allocated to various space radiocommunication services and the geostationary satellite orbit for these services." The resolution further noted that "the use of the allocated frequency bands and fixed positions in the geostationary satellite orbit by individual countries or groups of countries can start at various dates depending on requirements and readiness of technical facilities of countries."[24]

Therefore, it was resolved that "the registration with the ITU of frequency assignments for space radiocommunication services and their use should not provide any permanent priority for any individual country or groups of countries and should not create an obstacle to the establishment of space systems by other countries."[25] Thus, the regulation of the radio spectrum and orbit resources were linked by adoption of Resolution Spa 2-1.

Although supported by the Members in attendance, the foregoing resolution did not possess binding force. Still, some commentators have held that the language of the resolution "is a clear rejection of the view that a registration with the IFRB of a national assignment of a radio frequency would establish permanent priority for the registrant respecting a particular frequency." Further, registration of a space service frequency assignment "should not constitute a barrier to the establishment of space systems by other countries."[26]

The principle of equitable access as provided for in Resolution Spa 2-1 was later incorporated into Article 33 of the revised 1973 ITU Convention. (See Section 6.3 on the 1973 Plenipotentiary Conference.) In 1979, Resolution Spa 2-1 was replaced with the more strongly worded Resolution AY. (See Chapter 7 for coverage of WARC-79.) But in 1971 further support for this principle was also found in a recommendation that emanated from the Space Telecommunications Conference.

Recommendation Spa 2-1, entitled "Relating to the Examination by World Administrative Radio Conferences of the Situation with Regard to Occupation of the Frequency Spectrum in Space Radiocommunications," affirmed the proposition that nations should be able to establish their necessary telecommunication links and focused on the efficient and economic use of radio frequencies. Thus, the Administrative Council was invited to present for consideration at the next appropriate WARC all aspects of the use of the frequency bands concerned including those already registered with the IFRB.

Recommendation Spa 2-1 was enacted considering " . . . that despite the provisions of Article 9A of the Radio Regulations and the principles adopted by this Conference, which provide for full consultation and coordination between administrations with a view to the optimum accommodation of all space systems, it is possible that as the use of frequencies and orbital positions increases, administrations may encounter undue difficulty in one or more frequency bands in meeting their requirements for space radiocommunication; . . ."[27]

6.2.4 The Issue of DBS: Resolutions Spa 2-2 and Spa 2-3

As has already been mentioned, the broadcasting-satellite service was probably the most controversial topic considered at the WARC-ST. The use of direct broadcast satellites, or DBS, had previously been debated for some years in other international fora. No consensus agreement had been reached by 1971 on DBS. The debates generally concerned the principle of prior consent for broadcasts, either sound or television, *versus* the free flow of information.

The socialist and third world countries viewed DBS as a threat to their national sovereignty, a threat that had a potential for negative impact upon their cultural heritage. Consequently, these countries called for the incorporation of the principle of prior consent for broadcasts into international agreements on DBS. Western nations, the United States in particular, regarded prior consent as a mechanism connoting censorship which is incompatible with the US Bill of Rights and the inalienable human right of freedom of information.

No consensus was ever reached on the issue of prior consent, and this issue still remains at impasse. The issue was discussed to a point of near consensus, but the UN Declaration on DBS finally became the first space law resolution to be sent to the General Assembly from the UN Committee on the Peaceful Uses of Outer Space (COPUOS) without having attained consensus. The United States objected to the adoption of the DBS Declaration because of the absence of a clear and satisfactory resolution of the issue concerning prior consent, censorship, and information flow. (See Chapter 9 for a detailed discussion of DBS.)

Because DBS was regarded as a threat to national sovereignty and as a threat to the preservation of their cultural heritage by many socialist and third world states, they called for a regulatory means to limit such broadcasts — an *a priori* frequency allotment plan for the BSS. Many countries also saw the adoption of such a plan as a way to obtain their share of radio frequencies and as a way to circumvent the political impasse over prior consent.

When the 1971 Conference revised the regulations for coordination, notification, and registration of the space services sharing frequency bands with the terrestrial services, the title of Article 9A specifically excluded stations in the broadcasting-satellite service. This was an important indication of the still infant stage of development of this new service. The absence of coordination procedures and related guideposts for the BSS, however, was unsettling to many conference participants, many of whom feared that propaganda broadcasts could invade their sovereign territory without consent.

While many nations wanted to preclude the potential threat of television transmissions across their borders and others wanted to obtain their share of the radio resource, the European community wanted to plan the use of the spectrum so as to facilitate the implementation of a terrestrial microwave system at the same frequencies used by DBS.[28] As a consequence, the conference delegates agreed to fill the regulatory vacuum associated with the BSS by revising Article 7 of the Radio Regulations and by adopting Resolutions Spa 2-2 and Spa 2-3.

In ***Resolution Spa 2-2***, entitled "Relating to the Establishment of Agreements and Associated Plans for the Broadcasting-Satellite Service," the conference considered "that it is important to make the best possible use of the geostationary-satellite orbit and of the frequency bands allocated to the broadcasting-satellite service." It also considered that satellite broadcasting could create harmful interference over a large area of the earth, and that other services with allocations in the same band need to use the band before the BSS would be set up. Therefore, it was resolved in Spa 2-2 "that stations in the broadcasting-satellite service shall be established and operated in accordance with agreements and associated plans adopted by World or Regional Administrative Conferences, as the case may be, in which all the administrations concerned and the administrations whose services are liable to be affected may participate."[29] This resolution, calling for *a priori* planning, was a notable exception to the general regulatory regime.

Resolution Spa 2-2 also made provision for the interim period before such agreements and plans entered into force, stating that ". . . administrations and the IFRB shall apply the procedure contained in Resolution No. Spa 2-3."[30] The provisions in ***Resolution Spa 2-3*** established a moderately detailed coordination procedure between space stations in the BSS and terrestrial stations and between space systems of other administrations. It also established the conditions for notification, examination, and recording in the Master Register of assignments to space stations in the BSS.[31] Thus, this resolution actually moved the BSS in the regulatory direction of Article 9A and Appendix 1A, without as yet providing the full force of a binding procedure.

6.2.5 Revision of Article 7 of the Radio Regulations to Accommodate DBS

To complement Resolution Spa 2-2, a provision was adopted by the Conference that revised Article 7 of the Radio Regulations. The intention of this revision was to calm the fears of various nations with regard to the potential danger of unrestricted satellite broadcasting. ***Article 7 Paragraph No. 428A*** of the revised Radio Regulations provided that "[i]n devising the characteristics of a space station in the broadcasting-satellite service, all technical means available shall be used to reduce, to the maximum extent practicable, the radiation over the territory of other countries unless an agreement has been previously reached with such countries."[32]

The meaning of paragraph 428A has received several interpretations. Some countries see it as making prior consent for direct broadcasting mandatory. Others view it simply as a way of effecting technical coordination for the purpose of preventing interference, and they make the point that the ITU is a technical not a political forum, its approach being to control transmissions, not content. At any rate, paragraph 428A does not prohibit unavoidable communications through spillover transmissions. Therefore, it would appear that the intention was not to incorporate the principle of prior consent into the Radio Regulations but simply to try to establish some sort of technical framework for DBS and to limit as far as possible the potential for misuse of the as yet undeveloped technology until international agreements and plans could be established.

6.2.6 Overview of the Space Telecommunications Conference

The ITU made many important contributions toward the regulation of space communication at WARC-ST. A number of new space services were defined, thus substantially expanding Article 1 of the Radio Regulations. The conference added to the Table of Frequency Allocations bands for the newly established space services as well as revised allocations to other services. Most importantly, the 1971 Space Telecommunications Conference revised Article 9A of the Radio Regulations, introducing new provisions related to the station-keeping of satellites in the GSO, and initiated the process of long-range planning for the GSO.

Thus, the accomplishments of the WARC-ST were substantial, further extending the legal principle of nondiscriminatory, equitable access to the spectrum-orbit resource, a principle that had been enunciated in international fora in previous years. A turn in ITU policy was also initiated at the WARC-ST by calling for an *a priori* planning conference for the BSS. Before the end of the decade, Members of the ITU would see the 12 GHz frequency band planned for the BSS as well as a call for the planning of certain bands for the FSS — a clear indication that the orientation of the ITU on equitable access was moving toward a guarantee in practice from a guarantee in principle.

6.3 THE 1973 MÁLAGA-TORREMOLINOS PLENIPOTENTIARY CONFERENCE

6.3.1 General Provisions Amended

The International Telecommunication Convention enacted in 1973 at the Málaga-Torremolinos Plenipotentiary Conference attached great emphasis to the considerations enunciated in Resolution Spa 2-1 of 1971 and incorporated new and notable provisions for the regulation of space communication. The provisions related to space radiocommunication, as laid down in ***Article 33 of the International Telecommunication Convention*** — "Rational Use of the Radio Frequency Spectrum and of the Geostationary Satellite Orbit" — flowed directly from Resolution Spa 2-1.

Recall that Resolution Spa 2-1 recognized that the radio frequency spectrum and the geostationary satellite orbit were limited natural resources and emphasized efficient and equitable utilization of these resources. In 1973 this resolve was formally incorporated into the ITU treaty document and became binding upon all Members of the Union. Article 33 read as follows:

1. Members shall endeavour to limit the number of frequencies and the spectrum space used to the minimum essential to provide in a satisfactory manner the necessary services. To that end they shall endeavour to apply the latest technical advances as soon as possible.
2. In using frequency bands for space radio services Members shall bear in mind that radio frequencies and the geostationary satellite orbit are limited natural resources, that they must be used efficiently and economically so that countries or groups of countries may have equitable access to both in conformity with the provisions of the Radio Regulations according to their needs and the technical facilities at their disposal.[33]

Thus, in 1973 the radio frequency spectrum and the geostationary satellite orbit were internationally recognized as being integrally related for purposes of space communication. And while the principle of nondiscriminatory, equitable access to the spectrum-orbit resource had been enunciated in international fora for over a decade, it was not until 1973 that this principle became formal international communication law. At the same time, emphasis was placed on the efficient and economical use of the radio and orbit resources.

In 1973, just eight years after the last plenipotentiary conference, the majority membership of the ITU — the developing countries — succeeded in securing in principle equitable access to the spectrum-orbit resource. Within the next few years, the provisions of Article 33 of the International Telecommunication Convention would serve as the basis upon which these same countries would seek revision of the Article 9A coordination, notification, and registration procedures for the space services, claiming that these regulations were not in conformity with the principle of equitable access as provided for in Article 33.

In 1973, the International Telecommunication Convention was also amended to empower the IFRB to carry out greater regulatory responsibilities associated with space communication. Thereafter, ***Article 10*** of the Convention, entitled "International Frequency Registration Board," provided that the essential duties of the Board were:

(b) to effect . . . an orderly recording of the positions assigned by countries to geostationary satellites;
(c) to furnish advice to Members with a view to the . . . equitable, effective and economical use of the geostationary satellite orbit;
(d) to perform any additional duties, concerned with the assignment and utilization of frequencies and with the utilization of the geostationary satellite orbit, in accordance with the procedure provided for in the Radio Regulations, and as prescribed by a competent conference of the Union. . . .[34]

These provisions established various new functions for the ITU. The IFRB was now responsible for recording the assignments of geostationary satellite positions. While the 1965 Convention dealt only with the recording of radio frequency assignments, the 1973 Convention further charged the IFRB with the recording of satellite positions in the GSO. In establishing this new function, a direct legal association was made between frequency assignments and the orbital positions of geostationary satellites.

The ITU was also now charged with giving advice on the use of the GSO. This provision had the effect of moving the ITU from performing simply a ministerial function — that of recording assignments of satellites — to furnishing advice to Members, thus effecting policy related to the equitable, efficient, and economical use of the GSO.

6.3.2 Resolutions, Recommendations, and Opinions

Various resolutions were also passed in 1973 related to space communications. Pursuant to Resolution Spa 2-2 of 1971 and at the impetus of certain countries who asked the Administrative Council in the interim period between 1971 and 1973 to consider holding a regional administrative conference for the planning of the broadcasting-satellite service, ***Resolution 27*** was adopted by the Málaga-Torremolinos Conference. Under this resolution, the plenipotentiary conference decided to hold a world administrative radio conference no later than April 1977 for the planning of the broadcasting-satellite service in the frequency band 11.7-12.2 GHz in Regions 2 and 3 and 11.7-12.5 GHz in Region 1. The Administrative Council was instructed to make preparations for convening the conference. The planning of the BSS was to be done on a worldwide basis, rather than on a regional basis.[35]

The agenda of the conference called for the planning of the BSS and the establishment of the sharing criteria between the BSS and the terrestrial services to which the bands were allocated — the fixed service, the mobile service, except the aeronautical mobile, the broadcasting service, and the FSS in Region 2.[36]

The 1973 Conference authorized the planning of the BSS only in the 12 GHz band rather than in all the bands allocated in 1971 for this service. This was done primarily because the use of the lower frequency bands allocated to the BSS — the 620-790 and 2500-2690 MHz bands — were already extensively used in many countries for terrestrial broadcasting and other terrestrial services. Thus, the Radio Regulations placed severe restrictions to avoid interference upon the use of these bands for satellite broadcasting, making it suitable for such broadcasting only in countries that did not use or plan to use the bands for terrestrial services.

While the lower frequency bands were used extensively for fixed and mobile terrestrial services in western Europe, making the possibilities for the development of the BSS quite restricted, the higher frequency bands allocated to the BSS were not in practical use yet. These bands were the 22.5-23, 41-43, and 84-86 GHz bands. Therefore, planning of the frequency bands for the BSS was concentrated on the highly desirable 12 GHz band, and the conference that was to plan this band was scheduled for 1977.

The 1973 Málaga-Torremolinos Conference also adopted a resolution that set the stage for WARC-79 — the first general administrative radio conference to deal with complete revision of the Radio Regulations since 1959. Rapid technical advances in the radio art, changes in various services, and a necessity to harmonize the decisions of various conferences that had taken place in the last twenty years gave impetus to the proposal for a general revision of the Radio Regulations. As a consequence ***Resolution 28*** was passed calling for the convening of a conference in 1979 to consider general revision of the Radio Regulations and instructing the Administrative Council to prepare for it.[37] The 1979 WARC, called for by the 1973 Plenipotentiary Conference, would prove to be the forum that paved the way for some of the most revolutionary changes in the regulatory structure of the ITU related to space communication.

Finally, various resolutions related to development assistance were adopted at the 1973 Conference. As has been noted, the developing countries were in the majority, and the concerns of these countries almost completely dominated the plenipotentiary conference. Ten major proposals related to development assistance were approved in plenary meeting, many of which were extensions of recommendations passed in Montreux.[38]

Resolution 19 requested that the ITU make a special effort to help the least developed countries meet their communication needs. ***Resolution 21***, the most controversial resolution pertaining to development assistance, called for a special technical cooperation fund to be set up to meet the needs of the developing countries who submitted urgent requests for assistance. A similar proposal had been put forth in 1965 at Montreux but was not passed. ***Opinion 2***, also approved at the 1973 Conference, asked the developed countries to consider requests made by the developing countries for favorable treatment in areas of communication.[39]

Also in 1973, the subject of creating ITU regional offices in the developing countries again came up. So did the proposal for establishing a new consultative committee for technical cooperation. Both proposals were defeated, but proposals for regional offices would surface again and be approved at the next plenipotentiary conference, instructing various organs of the ITU to carry out studies aimed at achieving a strengthened regional presence and instructing the following plenipotentiary conference to review such studies.

The role that the developing countries played in policy formation became stronger in the 1970s, especially with reference to development assistance. This aspect bears directly on space communication. Although in the developed countries development took place along traditional lines, with railroads, landlines, highways, airports, and then satellites being put into place, in the developing countries the pattern of development often is reversed. As the cost of traditional, ground-based infrastructure is prohibitive, given the nature of the economy of the lesser developed countries, these countries look to space communication as a more economical and efficient alternative that can quickly connect remote areas, even though, in many cases, these remote areas lack the telecommunication infrastructure needed to make the use of space communication systems economical.

Nevertheless, there are many good examples of this reversed pattern of development. Indonesia has set up the Palapa satellite network to connect its chain of many hundreds of islands. It would have taken many years and billions of dollars to connect this large and dispersed nation of over 200 million people. Satellite communications made this connection possible at a fraction of the cost and time. A different but equally descriptive example of how satellite technology has enabled lesser developed countries to modernize is the case of Brazil and its Brazilsat. Although Brazil does not have the problem of trying to connect hundreds of islands, it is a massive country with many remote areas like the Amazonian region. Again, development is taking place in reversed sequence.

It is not difficult to see why the developing countries have such a great interest in development assistance, particularly with reference to space communication. Many countries look to space communication as their way into the twenty-first century — a way to carry them out of their preindustrial state directly into the postindustrial era. By the end of the decade of the 1970s, the developing countries had obtained a major influence on directing policy formation in the areas of development assistance and space communication.

6.4 THE 1977 WORLD ADMINISTRATIVE RADIO CONFERENCE FOR THE PLANNING OF THE BROADCASTING-SATELLITE SERVICE

On 10 January 1977, 660 delegates from 113 countries met in Geneva for the World Administrative Radio Conference for the Planning of the Broadcasting-Satellite Service in Frequency Bands 11.7-12.2 GHz (in Regions 2 and 3) and 11.7-12.5 GHz (in Region 1). When this conference (WARC-BS or WARC-77) closed five weeks later on 13 February 1977, plans for the broadcasting-satellite service in Region 1 (Africa, Europe, the Middle East, the USSR, and Mongolia) and Region 3 (Asia and the Pacific) had been adopted, including provisions for future plans that would govern Region 2 (the Americas and Greenland).[40]

The adoption of an *a priori* method of planning, which relies upon competent conferences to divide up the spectrum and assign radio channels and orbital positions to the nations of the world before they have the need or ability to use them, marked another step toward a break in ITU tradition. With the signing of the Final Acts and the adoption of a plan in which the interests of all countries had been determined in advance, an attempt was made to replace the first-come, first-served practice of obtaining international recognition for use of the spectrum-orbit resource, at least for broadcasting satellites operating in the 12 GHz band.

In view of the inherent political, economic, social, and cultural implications of the broadcasting satellite technology, the conference proceeded to fashion a plan for the BSS that would satisfy the mandate of resolutions and recommendations that had been put forth in earlier conferences. The Preamble to the Final Acts of WARC-BS reiterated principles that had been enunciated in previous conferences and reflected the two major concerns of the conference — efficient and equitable use of the radio and orbit resources. These considerations, incorporated into the Preamble of the Final Acts of the WARC-BS, read as follows:

> — bearing in mind the importance of making the best possible use of the radio-frequency spectrum and the geostationary-satellite orbit as well as the need for an orderly development of the services to which these bands are allocated;
> — taking into account the equal rights of all countries, large and small, even those countries which are not represented at the Conference; . . .[41]

Thus, when the provisions of the Final Acts of the WARC-BS entered into force on 1 January 1979,[42] the principles of efficient and equitable use of radio frequencies and orbital positions for the BSS in the 12 GHz band were interpreted to mean *a priori* assignment of the spectrum-orbit resource among all Members of the Union in Regions 1 and 3. It was decided at the conference that a plan for Region 2 would be deferred until a future conference.

6.4.1 Agenda of the Conference

The Broadcasting-Satellite Conference, concerned as it was with the planning of the BSS in the 12 GHz band, was faced with the international regulation of a service for which no routine experience yet existed. The establishment of a plan for the broadcasting-satellite service was, however, facilitated by the fact that all the physical properties of radio related to this service, such as frequency, field strength, time, orbital position, and wave polarization, could be taken into account because broadcasting systems use homogeneous technical standards and operational configurations over long periods of time. As a consequence, fairly simple and rigid rules for the broadcasting-satellite service could be established.[43]

The conference had three major mandates:

1. to establish the sharing criteria for the bands 11.7-12.2 GHz (in Regions 2 and 3) and 11.7-12.5 GHz (in Region 1) between the broadcasting-satellite service and the other services to which these bands were allocated, namely: the fixed service; the mobile service (except the aeronautical mobile); the broadcasting service; and the fixed-satellite service (Space-to-Earth) in Region 2;
2. to plan for the BSS in those bands keeping in mind the principle of efficient and equitable use as enunciated in Resolution 27 of 1973 and Resolution Spa 2-2 of 1971; and
3. to establish procedures governing the use of these bands by the BSS and by the above mentioned services to the extent considered necessary by the conference.[44]

To prepare for the 1977 Conference, regional seminars were held to familiarize the administrations concerned with the various factors that needed to be considered as the conference progressed. Two of these seminars were held under the sponsorship of the ITU, one in Kyoto, Japan, and one in Khartoum, Sudan. Altogether, more than 140 delegates from 62 countries participated in these two seminars.[45]

Structured like any other radio conference, with committees such as steering, credentials, and budget control, the conference proceeded to discharge its duties. The main part of the work of the conference was handled in Committees 4, 5, and 6 — the Technical, Planning, and Procedures Committees. Drawing upon technical studies conducted over the past fifteen years by the CCIR and from a report compiled by the IFRB of all the information sent by administrations regarding their foreseeable requirements for the BSS, the conference was able to decide upon a plan for the BSS in the 12 GHz band.

At the beginning of the conference, it was decided that its conclusions would be presented in the form of Final Acts, containing a plan, procedures, annexes, protocols, resolutions, and recommendations. It was also decided that the Final Acts should be incorporated into the Radio Regulations at the next general WARC that was to be held in 1979.[46] After the Final Acts had been incorporated into the Regulations, they would be binding on all Members.

6.4.2 The Broadcasting-Satellite Plan

The 1977 Broadcasting-Satellite Plan alloted the radio resource roughly proportional to the geographic size of a country. Specific orbital positions, broadcasting channels, and geographic areas of coverage were assigned to countries, except those in the western hemisphere. The Plan resulted in a 42-page table of computer calculations presenting a collection of the technical parameters involved. As drawn up, the Plan contained the following characteristics: country symbol and IFRB serial number, nominal orbital position in degrees, frequency channel number, boresight geographic coordinates, antenna beamwidth, orientation of the ellipse, polarization, and effective isotropic radiated power (EIRP).[47]

Planning for Regions 1 and 3 was carried out on the basis of this technical data. Extensive computer data processing of the technical information took place to obtain a plan that could be considered optimal in terms of efficient and equitable use of the spectrum-orbit resource, as was called for in Article 33 of the 1973 International Telecommunication Convention.[48]

It was the province of the Technical Committee to establish the list of technical parameters, which essentially assigned to specific countries the frequency and orbital positions they could use for satellite broadcasting. Countries were alloted from 2 channels (Brunei) to 65 channels (USSR) assignments, depending on their size, population and foreseeable communication needs.[49]

On the average, five channels were alloted to each service area in Region 1 and about four channels to service areas in Region 3, with the direction, size, shape, polarization, and power of each antenna beam and the technical characteristics of the transmission specified. As the available bandwidth was greater in Region 1, 40 channels were alloted to this region, while Region 3 received 25 channels. Region 1 and Region 3 were alloted satellite positions between 37 degrees west longitude and 146 degrees east longitude, with each position separated by 6 degrees.[50]

Explicit in the Plan, which is now contained in *Appendix 30 of the 1982 Radio Regulations*, was the indication of the nominal orbital position. This referred to the longitude of a position in the geostationary satellite orbit associated with a frequency assignment to a space station in a space radiocommunication service. The position is measured in degrees east or west from the Greenwich Meridian.[51] (See Figure 6.1.)

Thus, the Plan for the BSS in the 12 GHz band for Regions 1 and 3 stipulated a specific frequency and orbital position to be used in conjunction with each other for satellite broadcasting. Note, however, that the Plan covered only downlinks in this service and not uplinks, or so-called "feeder links." This issue would be raised at WARC-79, with various countries

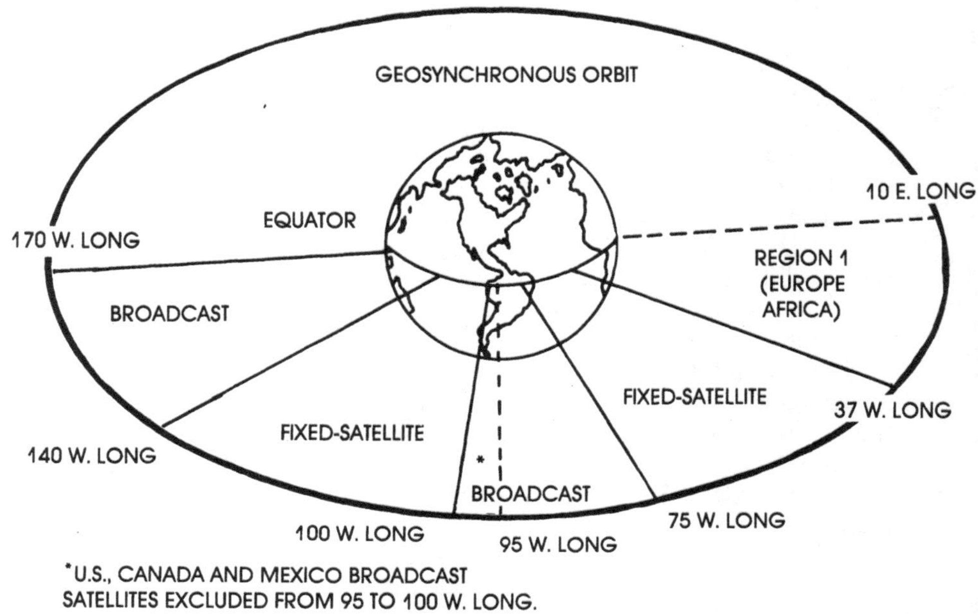

Figure 6.1 Arc Segmentation Plan Adopted by the 1977 World Administrative Radio Conference on the Broadcasting Satellite Service

Source: *Aviation Week and Space Technology*, July 23, 1979. Reprinted with permission.

calling for the planning of the feeder links.

With reference to the coverage of assignments in the Plan, Article 7, No. 428A of the Radio Regulations as revised in 1971 was applied. Recall that this provision stipulated that all technical means available were to be used to reduce as far as possible any spillover of signals into other countries with which no previous agreement had been reached. Thus, according to the Plan adopted at the 1977 WARC-BS, the BSS in the 12 GHz band was to be used only for domestic broadcasting.[52] Spillover was reduced to a minimum consistent with No. 428A of the Radio Regulations.

Nevertheless, the 1977 Plan for the BSS in Regions 1 and 3 has proven to be limited in terms of flexibility. When multinational systems began to be discussed and planned, it was found that the 1977 Plan could not accommodate selective revision for the benefit of a few countries without doing violence to the established plan. This rigidity was forecasted by the

United States and was the main reason the US would not accept the 1977 Plan. The US opposition led to postponement of planning for Region 2. Hence, the Plan as adopted in 1977 resulted in a system in which DBS was to be used only for national broadcasting and not international service. Several "regional system" proposals have been blocked and frustrated by the 1977 Plan to the aggravation of many of the administrations which most supported the Plan. "They made the Plan before they knew what they wanted and the Plan significantly constrained their choices."[53]

6.4.3 Planning for the Americas Deferred

In Region 2, the Americas and Greenland, the frequency band 11.7-12.2 GHz is shared with various services, including the fixed-satellite service. As a result, a serious problem of future interference existed in Region 2 relating to the BSS, with the high-power flux density of the broadcasting signal potentially able to interfere with the ability of earth stations in the FSS to receive clear signals. In addition, most administrations in Region 2, in general agreement with the United States, were reluctant "to accept forthwith a Region 2 broadcasting plan which might have crystallized a *de facto* situation that did not ensure optimal use of the spectrum-orbit resource in the band concerned."[54]

Accordingly, an interim arc segmentation plan was adopted by the 1977 Conference that allocated separate orbit segments to each service, and Article 12 of the Plan established provisions governing the BSS in Region 2 until a detailed plan for this region could be adopted.[55]

The WARC-BS decided that the planning of the BSS in Region 2, as well as necessary planning for the FSS, should be undertaken by a future conference. Consequently, *Resolution Sat-8* was adopted. The resolution called for a regional administrative radio conference to be convened for the detailed planning of the space services in the frequency band 11.7-12.2 GHz in Region 2.[56]

Resolution Sat-9 of the Final Acts of WARC-BS identified the goal for the future regional conference, eventually held in 1983, to be the drawing up of a detailed plan for the spectrum-orbit resource. The plan would provide for the detailed assignment of orbit positions and available frequency channels, "ensuring that the broadcasting-satellite service requirements of the various administrations are met in an equitable manner satisfactory to all the countries concerned."[57]

Recommendation Sat-8 further elaborated on the task of the Region 2 broadcast-satellite conference of equitably planning the BSS. It stated that the plan for Region 2 must guarantee to each administration "a minimum number of channels (4) for the operation of the broadcasting-satellite service. Above this minimum, the special characteristics of the countries (size, time zones, language differences, etc.) shall be taken into account."[58] Eventually, Resolutions Sat-8 and Sat-9 and Recommendation Sat-8 were replaced with Resolution 701 of the 1979 General World Administrative Radio Conference.

Thus, the matter of planning the BSS in the 12 GHz band for Region 2 was deferred to a later date. Just as the planning of the BSS in the 12 GHz band for Regions 1 and 3 was nominally predicated upon adherence to the principles of efficient and equitable use of the radio resource and the GSO, so also was to be the plan for Region 2.

The United States objected to the 1977 Plan. Objection, however, was not based on a lack of concern for equity. It was the position of the United States that a proper choice of technical parameters, implemented in a regulatory regime, would ensure that the requirements of all countries for BSS use of the GSO could be met. The US position was that it is in the interest of all countries to make provision for innovations in technology and changing patterns of space communication system uses over time. Consequently, the United States has insisted that ITU regulations for space services and for use of the GSO must contain suitably flexible procedures, appropriate technical factors, and decision-making mechanisms that would allow each country to obtain on a timely basis communication satellite services that meet that country's needs. The United States has consistently recognized that all administrations have equal right of access to the orbital and spectrum resources necessary to meet their requirements.[59]

6.4.4 Coordination, Notification, and Registration of Frequency Assignments

The work of Committee 6 (Procedures) of the 1977 Conference, set forth the regulatory system for the BSS. Recall that coordination procedures were not established for the broadcasting-satellite service in 1971. Mandatory provisions relating to coordination, notification, and registration of assignments in the BSS were established at the 1977 Conference and agreed to, because a detailed plan for this service had been agreed to.

Unlike the regulatory regime established for the space services sharing with equal rights frequency bands above one GHz with the terrestrial services, the regime established for the BSS in Regions 1 and 3 did not allow the assignment of frequencies for broadcasting-satellite services, terrestrial services, or other space services that could cause interference to service in the agreed Plan without the consent of the IFRB and all countries concerned.[60] Thus, the mandatory regime established by the WARC-BS was not limited to assignments in the BSS.

Article 6 of the 1977 Regulations dealt with the coordination, notification, and recordation in the Master International Frequency Register of frequency assignments to terrestrial stations affecting allocated frequencies to the BSS that had been agreed to in the Plan. The regulatory procedures were to be applied if the necessary bandwidths of the two transmissions overlapped or if the power flux density produced by the proposed terrestrial transmitting station exceeded a specific value as derived in Annex 3 of the Plan at one or more points on the edge of the service area that was within the coverage area of the broadcasting-satellite station.

Any administration assigning frequencies to terrestrial services or other space services that involved frequencies in the Plan for the BSS was required to coordinate the assignment by following the coordination procedures established for the BSS. If coordination were not successfully completed, the assignment was not to be recorded in the Master Register, and the administration was not to bring the service into operation until coordination was successfully completed. The administration could, however, upon agreement of the IFRB and affected administrations, put the service into use for a specified period of time. Although an assignment could be validly recorded for a temporary, specified period of time, such use did not allow for any subsequent priority or the pre-empting of a later service brought into use in accordance with the Plan.[61]

Under the 1977 regulatory regime for the BSS, administrations had to notify the IFRB not earlier than five years and not later than 90 days before bringing into use a service using a frequency allocated in the Plan. The IFRB then examined the notice with respect to conformity with the Convention, the Radio Regulations, and with the Plan. If the finding was favorable, the assignment would be recorded in the Register. If the finding was unfavorable, it was returned to the submitting administration with the IFRB's recommendations. If then the notice were resubmitted without modification or with modification that still resulted in an unfavorable finding, the administration was obligated not to bring the assignment into use until such time as it received a favorable finding from the IFRB.[62] Thus, by establishing a mandatory regulatory regime for the BSS from which deviations or modifications were not allowed except with the express consent of all affected administrations and the IFRB, the broadcasting-satellite service acquired absolute protection.

6.4.5 Efficiency Sacrificed for Equity

The Plan for the Broadcasting-Satellite Service as adopted in 1977 was intended to meet the requirements of the BSS in the planned bands for at least fifteen years from the date of the entry into force of the Final Acts of the conference. The Plan, however, has been characterized as sacrificing the efficient use of the spectrum-orbit resource in order to achieve equitable access to that resource. Instead of requiring the late-comer to seek accommodation, the superior position rests with the nation whose service conforms to the Plan, even if the administration has no intention or ability to use the frequencies and orbital positions in the near future and regardless of how long a nonconforming service has been in use.[63] In mid-1987, a decade after the Plan was adopted, there were still not operational BSS systems established in Europe or Africa. Only the USSR and Japan broadcast by satellite.

Those states that oppose *a priori* planning of the spectrum-orbit resource feel that technology can render such plans obsolete and could consequently result in a less efficient utilization of the resource. The main criticism of allotment plans is that they are rigid, lacking the flexibility to adjust quickly and easily to changes in requirements and in technology. This is especially true where radio technology is concerned, as it changes so rapidly, particularly with reference to space radiocommunication.

For example, since adoption of the 1977 Broadcasting-Satellite Plan, technological advances have allowed for ever closer spacing between satellites, while retaining protection against interference. The 1977 Plan, however, locked in satellite separation at six degrees for the BSS. Thus, the long-term planning that was engaged in at the 1977 Conference, based as it was upon certain technical standards of the time, has since been made obsolete and has never been used. Yet, the countries of Regions 1 and 3 are bound by such agreement until such time that it is changed.

Also, the adopted interim arc segmentation plan that established separate orbit segments for the FSS and the BSS in Region 2 was later found to put severe and unnecessary constraints upon the number of satellites in both services that could use the arc.[64] As a consequence, WARC-79 abandoned the arc segmentation plan and adopted a frequency segmentation plan for Region 2.

An additional consideration concerning efficiency sacrificed for equity relates to the use of space platforms with multiservice or multifrequency telecommunication capabilities. The use of multipurpose platforms for telecommunication may introduce constraints to the harmonization process, especially if the orbital position of the satellite has been determined by a previous plan.

6.5 CONCLUSION

The 1970s witnessed great advancements in the use and regulation of international satellite communication. (See Table 6.2.) At WARC-ST-71 frequencies were allocated to new services, such as the FSS, and to redefined services, such as the BSS. Article 9A of the Radio Regulations was revised concerning notification, coordination, and registration. The Conference provided for frequency sharing, for equal rights and no permanent priority to the spectrum-orbit resource, for economy and efficiency in the use of the resource, and for the use of all technical means to the maximum extent practicable to reduce transborder broadcasts that had not been agreed to between countries. WARC-71 also exempted the BSS from Article 9A and called for the planning of the BSS by a future conference.

In 1973, the Málaga-Torremolinos Plenipotentiary Conference became the first to place the principle of equitable access to the spectrum-orbit resource into the International Telecommunication Convention in Article 33. This article declared the spectrum-orbit resource to be a "limited natural resource" and also reaffirmed the traditional principles of efficiency and economy for the management of the resource. Article 10 of the 1973 Convention gave the IFRB additional responsibilities in the space field, including responsibility for the recording of national assignments to positions in the GSO.

Yet, despite the promise of space communication regulatory developments during the 1970s, the rigid international planning process adopted by WARC-BS-77 caused dissatisfaction, especially to developed countries that are bringing new technology into use — technology that outstrips the assumptions embedded in the 1977 Plan. It became clear that simplistic or rigid assumptions and provisions, especially in treaty documents, were counterproductive in spectrum-orbit management.

Indeed, it has been maintained that the term "spectrum-orbit resource" is deceptively over-simplified, that the resource is in reality:

> . . . a complex construct describing the physical properties associated with the radio energy used for transmissions. . . . field strength, time, spatial volume . . . and wave polarization. . . . each one of these five different dimensions is capable of virtually infinite subdivision and combination. . . .[65]

This reality was perhaps not clearly seen by most ITU Members as the time approached for the long awaited 1979 GWARC. Unquestionably, the very large increase in membership from the developing world had moved the focus of the WARC-79 toward equitable access in general and planning procedures for satellite communication in particular. Nevertheless, the

Table 6.2 Legal Principles and Provisions for Space Communication—1971 to 1977*

1971 WORLD ADMINISTRATIVE RADIO CONFERENCE FOR SPACE TELECOMMUNICATIONS	
Article 1	Fixed-Satellite Service and Mobile-Satellite Service replace Communication-Satellite Service; BSS expanded; Earth-Exploration and Maritime-Mobile Satellite Services (MMSS) defined, and frequencies allocated for their use—MMSS facilitates establishment of INMARSAT.
Article 5	About 15,000 MHz of new spectrum allocated to space services.
Article 9A	Revised to include mandatory procedure of advance publication of information on planned satellite systems; to take place prior to coordination.
Article 7, Para. No. 428A	Obligation to minimize radiation overspill of BSS signals.
Appendix 1B	"Advance Publication Information to be Furnished for a Satellite Network" sets forth basic operational data to be provided IFRB on any planned satellite system.
Resolution Spa 2-1	*Recognized spectrum-orbit as natural resources, thus linking them; emphasized efficient and economical use; stated countries have equal rights to use spectrum-orbit resources.*
Recommendation Spa 2-1	*Affirmed proposition that countries should be able to establish their necessary telecommunication links, and focused on efficient and economical use of the radio spectrum.*
Resolution Spa 2-2	Called for *a priori* planning of the BSS.
Resolution Spa 2-3	Established detailed coordination procedure for BSS with conditions for notification, examination, and registration in the IFRB Master Register of BSS assignments.

Table 6.2 (cont'd)

1973 MÁLAGA-TORREMOLINOS PLENIPOTENTIARY CONFERENCE	
Article 33	*Principle of nondiscriminatory equitable access to spectrum-orbit resource established as formal international law with emphasis on efficient and economical use.*
Article 10	Amended to charge IFRB with greater regulatory responsibilities for space communication, including recording of satellite positions in GSO and advising Members on its efficient and economical use.
Resolution 27	Called for WARC-BS to plan BSS.
Resolution 28	Called for WARC-79 to consider general revision of Radio Regulations.
Technical Cooperation	Ten Resolutions including:
Resolution 19	Special measures for least developed countries to meet their communication needs;
Resolution 21	Special fund for technical cooperation for developing countries;
Opinion 2	Favorable treatment in areas of communication for developing countries.
1977 WORLD ADMINISTRATIVE RADIO CONFERENCE FOR THE PLANNING OF THE BROADCASTING-SATELLITE SERVICE	
Article 11	Established *a priori* allotment Plan for the BSS in Regions 1 and 3.
Resolution Sat-8	Called for RARC for equitable planning of BSS in Region 2.
Resolution Sat-9	Identified goal of future RARC as drafting of plan for BSS.
Recommendation Sat-8	Elaborated on equitable planning of Region 2 BSS by identifying minimum number of frequency channels for each administration.

**Note*: Italics refer to legal principles concerning equity.

developed world continued to express the expectation and desire that the ITU use a more flexible and appropriate method to overcome the disadvantages of a rigid plan such as that of 1977, while at the same time guaranteeing equitable access to the space segment for all nations without sacrificing efficient and effective use of the spectrum-orbit resource. An attempt to achieve this delicate balance is the subject of the next chapter.

REFERENCES

1. *ITU Administrative Council Resolution 653, May, 1969.*
2. Jasentuliyana, Nandasiri, "Regulations Governing Space Telecommunication," *Manual on Space Law,* Vol. 1, Dobbs Ferry, NY: Oceana Publications, 1979, pp. 195-238, at 227.
3. *Final Acts World Administrative Radio Conference for Space Telecommunications, Geneva, 1971,* (herein cited as *Final Acts WARC-ST*).
4. Wallenstein, Gerd D., "I.T.U. Case Studies," in *International Telecommunication Agreements,* Dobbs Ferry, NY: Oceana Publications, 1979, p. 11.
5. *Final Acts WARC-ST,* Art. 1.
6. *Ibid.*, Para. 84AP Spa 2.
7. *Ibid.*, Para. 84AP.1 Spa 2.
8. *Ibid.*, Para. 84APA Spa 2.
9. *Ibid.*, Para. 84APB Spa 2.
10. Busak, Jan, "Space Telecommunications at Present and in Future," *Proceedings of the Twenty-Second Colloquium on the Law of Outer Space,* New York: AIAA, 1979, pp. 29-34, at 30.
11. *Final Acts WARC-ST,* Annex 3, Art. 5.
12. *Ibid.*
13. *Ibid.*, Art. 1, Annex 3, Art. 5.
14. *Ibid.*, Para. 84AGC Spa 2.
15. See *Report of the United States Delegation to the World Administrative Radio Conference for Space Telecommunications.*
16. *Final Acts WARC-ST,* Art. 9A, Para. 639AA Spa 2.
17. *ITU Radio Regulations, 1982,* Art. 11, Para. 1042.
18. *Final Acts WARC-ST,* Appendix 1B.
19. *Ibid.*, Appendix 29.
20. Colino, Richard R., "International Cooperation Between Communications Satellite Systems: An Overview of Current Practices and Future Prospects," *Journal of Space Law,* Vol. 5, 1977, pp. 65-92, at 71.
21. Wallenstein, *supra* note 4, at 16.

22. *Ibid.*
23. *Ibid.*, p. 17.
24. *Final Acts WARC-ST,* Resolution Spa 2-1.
25. *Ibid.*
26. Christol, Carl Q., "The International Telecommunication Union and the International Law of Outer Space," *Proceedings of the Twenty-Second Colloquium on the Law of Outer Space,* New York: AIAA, 1979, pp. 35-51, at 42.
27. *Final Acts WARC-ST,* Recommendation Spa 2-1.
28. Codding, George A. Jr., and Anthony M. Rutkowski, *The International Telecommunication Union in a Changing World,* Dedham, MA: 1982, p. 48.
29. *Final Acts WARC-ST,* Resolution Spa 2-2.
30. *Ibid.*
31. *Ibid.*, Resolution Spa 2-3.
32. *Ibid.*, Art. 7, Para. 428A Spa 2.
33. *International Telecommunication Convention, Málaga-Torremolinos, 1973,* Art. 33, (emphasis added) (herein cited as *Málaga-Torremolinos Convention*).
34. *Ibid.*, Art. 10.
35. *Ibid.*, Resolution 27.
36. See *Final Acts WARC-ST,* Annex 3, Art. 5.
37. *Málaga-Torremolinos Convention,* Resolution 28.
38. *Málaga-Torremolinos Convention.*
39. *Ibid.*, Resolutions 19 & 21, Opinion 2.
40. Mili, M., "World Administrative Radio Conference for the Planning of the Broadcasting-Satellite Service in Frequency Bands 11.7-12.2 GHz (in Regions 2 and 3) and 11.7-12.5 GHz (in Region 1)," *Proceedings of the Twentieth Colloquium on the Law of Outer Space,* New York: AIAA, 1977, pp. 346-367, at 351.
41. *Final Acts of the World Administrative Radio Conference for the Planning of the Broadcasting-Satellite Service in Frequency Bands 11.7-12.2 GHz (in Regions 2 and 3) and 11.7-12.5 GHz (in Region 1), Geneva, 1977,* (herein cited as *Final Acts WARC-BS*).
42. *Ibid.*
43. Rutkowski, Anthony M., "Space WARC: The Stake of Developing Countries," *Space Policy,* Vol. 1, 1985, pp. 240-243, at 241.
44. *ITU Administrative Council Resolution 762, 1975.*
45. Mili, *supra* note 40, at 350.
46. *Ibid.*, p. 352.
47. *Final Acts WARC-BS,* Art. 11.
48. Mili, *supra* note 40, at 356.

49. *Final Acts WARC-BS,* Art. 11.
50. *Ibid.*
51. *Ibid.*, Annex 8.
52. *Final Acts WARC-BS.*
53. Correspondence to authors from Stephen E. Doyle, Esq., chairman of the FCC Space WARC Advisory Committee for the ITU World Administrative Radio Conference on the Use of the Geostationary-Satellite Orbit and the Space Services Utilizing It, June 16, 1987.
54. Mili, *supra* note 40, at 348.
55. *Final Acts WARC-BS.*
56. *Ibid.*, Resolution Sat-8.
57. *Ibid.*, Resolution Sat-9.
58. Recommendation Sat-8.
59. See *Report of the United States Delegation to the World Administrative Radio Conference for the Planning of the Broadcasting-Satellite Service in Frequency Bands 11.7-12.2 GHz (In Regions 2 and 3) and 11.7-12.5 GHz (In Region 1).*
60. *Final Acts WARC-BS,* Arts. 5 & 6.
61. *Ibid.*, Art. 6.
62. *Ibid.*
63. *Final Acts WARC-BS.*
64. Robinson, Glen O., "Regulating International Airwaves: The 1979 WARC," *Virginia Journal of International Law,* Vol. 21, 1980, pp. 1-54, at 25.
65. Rutkowski, *supra* note 43, at 241.

Chapter 7

ITU Conferences: 1979-1983 — From Principle to Practice

7.1 INTRODUCTION

By the end of the 1970s, many significant changes had taken place in the world environment for telecommunication, especially with respect to space communication. At the same time as the demand for communication services was sharply increasing, the economically attractive parts of the radio spectrum were becoming congested. By the end of the decade also, more than two-thirds of the membership of the ITU was composed of developing countries. This balance of membership resulted in significant changes in ITU policy orientation, especially in regard to space communication.

The policy formulation in the ITU relating to satellite communication followed a trend established in the early 1960s, emphasizing nondiscriminatory, equitable access to the space segment. In Chapter 5 we traced evidence of this principle to UN General Assembly resolutions in 1961 and in 1962 and showed that these resolutions also could be found reflected in ITU Recommendation 10A as far back as 1963. Then in 1965, Resolution 24 of the Montreux Plenipotentiary Conference called upon all Members to recognize that " . . . all countries should have equal opportunity to use space telecommunication facilities."[1]

In Chapter 6 we saw that this principle of nondiscriminatory, equitable access was furthered in 1971 with the adoption of Resolution Spa 2-1, which stated "that all countries have equal rights in the use of both the radio frequencies allocated to various space radiocommunication services and the geostationary satellite orbit for these services."[2] Then in 1973 at the Málaga-Torremolinos Plenipotentiary Conference, Article 33 was adopted, emphasizing efficient and economical use of the radio spectrum and geostationary satellite orbit "so that countries or groups of countries may have equitable access to both . . . according to their needs and the technical facilities at their disposal."[3]

Throughout the 1970s, the formulation of satellite communication policy embraced the principle of nondiscriminatory, equitable access to the spectrum-orbit resource. The translation of this principle into practice, as interpreted by the majority of ITU Members, entailed the use of *a priori* planning of the radio frequency spectrum and the GSO. In 1977, a major break with the traditional regulatory regime occurred in the ITU, with the case by case, or *a posteriori*, method of securing recognition of use of the radio spectrum and GSO being supplanted by planning of the 12 GHz band for the broadcasting-satellite service in Regions 1 and 3. The emphasis had shifted from first-come, first-served use of the spectrum-orbit resource to the coordinated planning of it.

This change in regulatory focus, where an *a posteriori* method of rights vesting or securing recognition was displaced by an *a priori* scheme, came about largely as a result of the efforts of the developing countries, which wanted to secure spectrum-orbit rights for their future communication needs. The fears of the developing countries of being pre-empted from using the space environment to meet their needs served as the force behind the shift in policy orientation toward planning. Many states, however, in particular the United States, opposed the use of rigid planning to govern the allocations and use of the radio spectrum and the geostationary satellite orbit.[4]

This debate indicated that basic differences exist between the developed and the developing world with regard to spectrum-orbit resource management. The developing countries look to planning of this resource to ensure that their future communication needs can be met, while the developed countries see planning as a severe handicap placed upon technological innovation. As the disparity between states in their ability to use the spectrum has grown, so has the disagreement over the allocation and use of specific frequency bands for specific services. This is particularly true with respect to frequency bands for space radiocommunication services.[5]

Satellite services are a major area of great, global interest and controversy. The reasons for the controversy over satellite communication are understandable as satellite communication has become pervasive over the last fifteen years, emerging to become a central component in international and domestic communication systems. Central to the utility and the economic advantages of satellite systems is their connectivity, or ability to connect a great number of earth stations with each other through a single satellite. Thus, many states believe that many of their future communication needs can be met by geostationary satellite systems, the use of which could aid in reducing the need to establish as costly a terrestrial communication infrastructure as might otherwise be required. Many developing countries believe the use of satellites will help them leapfrog many problems that would be more difficult to solve in other ways.[6]

Of course, satellites do not entirely replace infrastructure on the ground, rather they complement and extend it. To make a BSS system make sense, for example, there has to be a population of receivers and program production sources. To make an FSS system make sense, there must be earth stations with associated display or redistribution capacity like telephones, or telex and data systems on the ground. The space segment of a national educational BSS system for Argentina, for example, was determined to cost only about 10 percent of the total system cost. Schools, teachers, books, pencils, paper, and ground equipment for reception and display of satellite signals were the bulk of the cost. India's Satellite Instructional Television Experiment (SITE) in the 1970s conclusively showed that good program production costs alone are many times the satellite costs. Satellites can do very little on their own and thus could be more of a problem than a solution.[7]

Nevertheless, in the late 1970s the clear preference of the majority of countries in the ITU was for some form of planning for the satellite communication services. The developing countries viewed this regulatory approach as being the most effective way of ensuring the realization of equitable access to the space segment and securing their fair share of the spectrum-orbit resource — an evolution from principle to practice.

7.2 THE 1979 WORLD ADMINISTRATIVE RADIO CONFERENCE

The 1979 World Administrative Radio Conference that convened in Geneva for 10 weeks and adjourned 6 December 1979 was the first general administrative radio conference since WARC-59. When the Final Acts of the 1979 WARC entered into force on 1 January 1982,[8] the 1959 Radio Regulations, as partially revised by subsequent specialized radio conferences, were superseded.

Because of the broad scope of its agenda, which included all of the major arrangements related to radio, GWARC-79 was of special importance. Unlike specialized or regional ITU conferences that have limited jurisdiction over only portions of the Radio Regulations, the 1979 Conference had general jurisdiction over all the Regulations. As a general conference, WARC-79 was held to reach global agreements on the international arrangements necessary for efficient, interference-free radiocommunication.

The large increase in membership of the ITU since 1959, mostly comprised of developing countries, made WARC-79 different in size and complexity from any previous general administrative radio conference. Membership in the ITU in 1979 totalled 154, with the developing countries constituting about two-thirds of this membership. One-half of the states eligible to attend WARC-79 did not even exist before 1959.[9] Concurrent with this increase in membership of developing countries there was a worldwide growth in demand for telecommunication services and a consequent demand for use of associated radio frequencies.

Other differences also distinguished WARC-79 from WARC-59. Conference sessions in 1979 were highly charged politically. In the world political arena of 1979, a confrontational north-south relationship existed. Sharply defined demands were made by the developing countries for a redistribution of the world's wealth and resources. These manifested at the 1979 Conference in demands for preferential treatment and guaranteed access to certain congested frequency bands and segments of the GSO. Countries that were still developing their telecommunication services wanted to prevent the developed countries from pre-empting their potential use of certain portions of the spectrum-orbit resource before they could determine their future needs.[10]

WARC-79 was also marked by a growing awareness among the developing countries of the power and the advantage they held over the developed countries — the advantage of their numbers and their willingness to threaten to use a one-nation, one-vote method of reaching agreement. Influence in the ITU, however, has not been primarily expressed through majority voting, although voting is provided for in the ITU Convention. The practice preferred by nearly all countries has been to reach decision by consensus. This has been especially true in the ITU, where the technical work must largely be done by consensus. Majority voting has been used mostly for political and procedural matters and in some committees. The potential for the use of majority voting in plenary sessions, however, may well have advanced the trend toward planning of the radio resource.

In widespread practice, of course, the actual use of majority voting could actually prove counterproductive because of the sovereign right of each country to take reservations and refuse to implement particular provisions of international agreements. If the large spectrum using states, for example, did not agree with decisions taken, they could choose not to implement the decisions. Consequently, a plan could fall under its own weight because of the extraordinary degree of cooperation required in the use of this resource. Thus, when the more than 1,600 delegates from 142 states met in Geneva in 1979 to consider nearly 17,000 proposals related to revision of the Radio Regulations,[11] the extensive work at hand was the subject of very emotional, political debate.[12]

7.2.1 Preparation for the Conference

Delegates to the conference did not come unprepared. Preconference activities alone had spanned many years. Recall that the convening of WARC-79 was initially called for in Resolution 28 of the 1973 Málaga-Torremolinos Plenipotentiary Conference. Then in 1976, the Administrative Conference set the agenda for the 1979 Conference. The agenda included the following:

- review and revision of the Radio Regulations adopted in 1959 and subsequently revised by specialized radio conferences, as far as the terminology, allocation of frequency bands, and directly associated regulations were concerned;
- review and revision of coordination, notification, and recordation provisions;
- consideration of Resolution Sat-10 of the 1977 WARC-BS concerning the possible rearrangement of the Radio Regulations;
- consideration of decisions adopted by various radio conferences held since 1959; and
- proposal to the Administrative Council and to the next plenipotentiary conference of a program for convening future radio conferences to deal with specific services like space radiocommunication.[13]

Also in preparation for the 1979 WARC, the Administrative Council in 1975 and in 1976 directed the CCIR and the IFRB to conduct preparatory activities. In 1977, the Council further charged the CCIR with arranging for a special joint meeting of CCIR Study Groups to act as a Special Preparatory Meeting (SPM) of the CCIR that would serve to provide the technical basis for the conference and for administrations in preparing their proposals. The SPM was held in 1978 and was attended by more than 700 delegates from 85 Member countries, 30 recognized private operating agencies, 18 international organizations, 10 scientific and industrial organizations, and three Specialized Agencies of the United Nations. Finally in 1979, three ITU regional preparatory seminars were held to discuss the results of the 1978 SJM and to gather information on and discuss regional matters.[14]

The preparatory work for WARC-79 was of direct benefit to the conference. The review of so large a number of proposals, however, challenged even the largest delegations. One conference participant noted that on each day the delegates of most delegations met to decide which meetings to attend.[15] The larger delegations had an advantage over the majority of smaller delegations because they could cover more meetings. The smaller delegations were hard pressed to cover even those meetings that were directly relevant to their situation. As a result, the conference decided not to hold concurrent meetings of the more important committees. Still, the major committees established subcommittees, which made it even harder for the smaller delegations to attend all the meetings.[16]

Committee 4, the Technical Regulations Committee, set up four subcommittees. Committee 5, the Frequency Allocations Committee, established six subcommittees, and Committee 6, the Regulatory Procedures Committee, established two subcommittees. In addition, the work of these subcommittees was farmed out to even smaller drafting groups. This system of distributing the work load taxed the smaller delegations.[17] As the conference turned its attention to revising almost the entirety of the Radio Regulations, however, the delegations from the developing countries, although small in size, were collectively large in number, widely distributed, and wielded great power.

7.2.2 Revision of the Radio Regulations

Thousands of proposals aimed at revising the Radio Regulations were considered at WARC-79. As is the case with each radio conference, many of these proposals were directed toward ways of finding frequency space for new uses and alteration of the Table of Frequency Allocations to accommodate changes from old uses. To incorporate the necessary changes adequately and as a result of advances in radio technology, WARC-79 extended the upper limit of the radio spectrum to 400 GHz from 275 GHz.

The 1979 Conference also eliminated the Additional Regulations, which had become part of the Radio Regulations in 1927 and served to separate so-called managerial provisions like rates from the more general principles of the Regulations. Recall that the United States had never signed the Additional Regulations because, as a government, the United States believed it could not be bound to regulations that were usually within the province of private interests.

Although WARC-79 was concerned with a myriad of radio issues and associated revisions, there were three areas of regulatory revision particularly related to space communication. These were revisions of the

Regulations pertaining to the use of the 12 GHz band in Region 2, planning of the fixed-satellite service, and planning of the feeder links for the broadcasting-satellite service.

7.2.2.1 The 12 GHz Band: Interim Arc Segmentation Plan for Region 2 Replaced by Frequency Segmentation Plan

In 1979 the World Administrative Radio Conference incorporated the Final Acts of the 1977 Conference and its associated Plan as an integral part of the Radio Regulations. The Plan, however, as adopted in 1977, contained a compromise provision on space broadcasting that resulted in an interim arc segmentation plan for Region 2. This interim provision was adopted to allay the fears of various Latin American countries that felt that extensive development of the fixed-satellite service could foreclose implementation of the broadcasting-satellite service before a detailed plan could be made for the region.[18]

Under the arc segmentation plan, two arcs in the geostationary satellite orbit were allocated for the BSS and two were allocated for the FSS for operation in the 11.7-12.2 GHz band. The arc for the western hemisphere extended from 37 degrees west longitude to 170 degrees west longitude. (See Chapter 6 and Figure 6.1.) Within this arc both the broadcasting and the fixed-satellite service shared the 12 GHz band.[19]

Subsequently, it was found that this segmentation plan put severe, unnecessary constraints upon the number of satellites that could use the arc. These contraints were experienced in both services as differences in the system designs created a poor frequency sharing situation between the FSS and the BSS. Also, under the arc segmentation plan, the FSS had no grandfather rights and was required to operate on a noninterference basis with the BSS.[20]

To arrive at a more efficient use of the geostationary satellite orbit and to resolve the difficult sharing situation in Region 2 between the FSS and the BSS, the United States proposed expanding the 12 GHz band and dividing it into separate frequency bands instead of designating separate segments of the orbital arc for the two services. The United States proposed to expand the 12 GHz band by 500 MHz, allocating the 11.7-12.2 GHz band to the FSS and the 12.2-12.7 GHz band to the BSS. This would allow each service to use the entire arc for the western hemisphere, while providing 500 MHz of spectrum for each service.[21]

Because of disagreement between the United States, Canada, and certain Latin American countries, the 1979 Conference ultimately adopted a modified version of the US proposal. The new regulations doubled the down-link bandwidth available for use by the FSS and the BSS in Region 2 and separated the two services. The FSS was allowed to operate in the 400 MHz band from 11.7-12.1 GHz, while the BSS could operate in the newly opened 400 MHz band from 12.3-12.7 GHz. Both services, however, had to share the 200 MHz band between 12.1-12.3 GHz. Thus, under the new regulations, the entire geostationary arc over Region 2 was opened to satellites of both types, and the frequency sharing between both services was limited to 200 MHz.[22] The issue of how ultimately to allocate the shared portion of the band, however, was left for resolution to a scheduled 1983 Regional Administrative Radio Conference (RARC) for the planning in Region 2 of the 12 GHz band.

Although the frequency sharing between the space services was largely eliminated as a result of the new regulations, these regulations also required the broadcasting-satellite service to share with the terrestrial fixed services, which included private microwave systems widely used in the United States. The new allocations for the BSS to share the 12.1-12.7 GHz band were a cause for concern to both private broadcasters and microwave users. Broadcasting interests looked at sharing as potentially resulting in interference to sensitive BSS earth station receivers operating in the same area as microwave fixed-station transmitters. The microwave users, on the other hand, felt that sharing with DBS was not feasible and opposed sharing in the 12 GHz band as they anticipated the burden of resolving any resulting interference problems would be placed on them. This concern was substantiated by a footnote to the Table of Frequency Allocations that placed terrestrial services on a noninterference basis with the BSS operating in accordance with a plan that was to be effected in 1983 for Region 2.[23]

The allocation of 200 MHz in the middle of the band, from 12.1-12.3 GHz, was reserved for use by both services until the 1983 Conference could finalize any decision. This decision was reached as a result of a compromise struck between the United States, Canada, and various Latin American countries including Brazil, Chile, Mexico, Venezuela, and Argentina. Canada felt that the US proposal to split in half the proposed 1000 MHz for the FSS and the BSS would interfere with its planned terrestrial microwave system. Canada also felt that such a distinct split would preclude the implementation of a hybrid satellite system operating in both the fixed and broadcast services. When it was pointed out that the 1977 arc segmentation plan also precluded such operation, Canada called for final resolution of the allocation issue to be put off until 1983 in order to study the various options. The various Latin American countries that were concerned with the issue chose sides with either of the two states. Finally a compromise was adopted that divided the band in two as the

United States had proposed, but provided special limited access for hybrid systems to the two separate bands in addition to the 200 MHz for shared use.[24]

The head of the US delegation to WARC-79 has characterized the intraregional debate as a north-south controversy, stating that Latin American support for the Canadian proposal expressed the "apprehension that somehow the US proposal would enable the United States to exploit the 12 GHz band before the Region could develop a specific allotment plan and protect the interests of all countries." To the frustration of the United States, which demonstrated how its proposal could increase the usuable arc by as much as three times, many states "were skeptical of US motives and seized on Canada's technical objection as a basis for advocating a postponement."[25] Thus, the north-south tensions that pervaded many international fora in the late 1970s also effected the formulation of space communication policy.

7.2.2.2 Planning of the Feeder Links for the Broadcasting-Satellite Service

The 1977 Plan for the broadcasting-satellite service in the 12 GHz band in Regions 1 and 3 covered only down-links; it did not include up-links, or feeder links. Feeder links are included in the fixed-satellite service so any fixed-satellite allocations could be used. An up-link is a transmission from the Earth to the satellite. A down-link transmission is from the satellite to the Earth. Because a communication satellite cannot use the same frequency to transmit and to receive at the same time without interfering with its own message, specific bands have to be designated for either up-link or down-link use. Although planning of up-link frequencies for the BSS was supported at WARC-79, the identification of the specific frequencies that would feed the BSS posed significant problems.

Led principally by India, the developing countries confronted the United States and other developed states with the proposal to place the feeder links in the 14.5-15.35 GHz band. These links were to be shared on a primary basis with terrestrial and mobile services. Strong objection to sharing of this band by broadcast satellite feeder links was expressed by the United States and a number of its allies because that specific band, especially the upper portion, was used for defense communications like radar. These states preferred an allocation for the feeder links in the 17 GHz range. After long and controversial debate, the conference adopted a compromise initiated by the US that allowed the use of several bands for up-link transmissions, at the option of individual administrations. By resolution, the conference also called for a future allotment plan to be developed by future conferences.[26]

The United States agreed to downgrade the status of radar from primary to secondary in the 17.3-17.7 GHz band so that allocations in this band to the broadcast satellite feeder links could enjoy primary status. This was a fallback position taken by the United States after it refused to downgrade the status of radar in the 14.5-15.35 GHz band. The United States, however, also agreed that over time it would move out of the lower band.[27]

Thus, after considerable controversy at WARC-79 over which allocations to make for the broadcast satellite feeder links — a controversy that threatened a major north-south political division — broadcast satellite feeder link allocations were finally made in three different bands through a negotiated, consensus settlement without voting. Besides the 14.5-14.8 GHz band being allocated for feeder links for countries outside Europe and for Malta, the other options made available were the 10.7-11.7 and the 17.3-18.1 GHz bands.[28] The latter band was expected to become the principal feeder link outside Africa and the Middle East.

7.2.2.3 The Fixed-Satellite Service

In recognition of past and projected growth of international satellite communication in the fixed-satellite service, the 1979 WARC substantially increased the bandwidth allocated to this service. The bandwidth allocated to the FSS in the 6/4 GHz bands was approximately doubled as was the bandwidth in the 14/11 GHz bands. This total allocation amounted to about 11 times the bandwidth of the previously allocated 6/4 GHz bands for use by FSS communication satellites. In addition to these bands, a bandwidth of 3,500 MHz remained allocated to the FSS in the 30/20 GHz band, a band where the next generation of domestic communication satellites is now under development. Additionally, more than 30 GHz was allocated to the FSS in the 50/40, 80/70, 105/95, 210/150, and 270/240 GHz bands. These bands, however, have not yet been used experimentally. WARC-79 also decreased the station-keeping tolerance for the FSS to plus or minus 0.1 degree to help facilitate reduced orbital spacing.[29]

The most controversial problems of the 1979 Conference pertained to the FSS. Generally, the disputes concerned whether new allocations to the FSS could be made below 10 GHz. This region of the spectrum below 10 GHz was shared and used extensively by the fixed-satellite service and by various terrestrial services. Allocations in the higher portions of the spectrum offered less congestion but could increase initial satellite development costs. Additional allocations below 10 GHz, however, could only be accommodated by displacing or crowding existing terrestrial services. Consensus on the issue was not easily reached.

Again led by India, several developing countries forwarded a proposal that called for downgrading radar in the 3.4-3.7 GHz band to a secondary status in order to accommodate fixed-satellite service by INTELSAT. This proposal was vehemently opposed by the United States and the United Kingdom, which used this band for national and allied defense radar systems.[30]

The 3 GHz band had long been shared equally by radars and fixed satellites in Regions 2 and 3. Because of the incompatibility of radar with the FSS, however, neither the United States nor INTELSAT had implemented the fixed-satellite service in this band because the United States used it for military radar operations. The proposal to downgrade radar to a secondary status was an attempt made by the developing countries to facilitate use of FSS systems, particularly by INTELSAT. Both the INTELSAT system and the US domestic system operate in the adjacent 3.7-4.2 GHz band. The United States proposed to maintain this status quo and to retain the primary status for radar in the 3.4-3.7 GHz band.[31]

The United States and its allies did agree to a provision urging countries to move radar out of the lower band when practicable. India demanded, however, that the date be specified when radar would lose its coequal status. Although the United States and its allies refused to accept this condition during subcommittee meetings, India's demand was generally accepted at the committee level.[32]

As a consequence, the United States refused to accept any decision to downgrade the radar status and threatened to effect a footnote to the Table of Frequency Allocations to exclude the FSS in the United States from operating in the 3.4-3.7 GHz band. This would in effect foreclose the use by INTELSAT of these frequencies for the fixed-satellite service because a major portion of INTELSAT traffic originates in, transits, or terminates in the United States.[33]

The threat of effecting a footnote to the Table of Frequency Allocations, however, had little impact on India's position, which the head of the US delegation has described as exhibiting only a "professed objective of helping INTELSAT," when really "India desired the frequencies primarily to accommodate her future domestic satellite needs." India had just come through a two-year coordination process over her domestic Insat satellite system and had experienced difficulty with INTELSAT and with the USSR in trying to get its domestic satellites coordinated. Other developing countries, however, ultimately saw that "their interest in providing for INTELSAT could not be forced over steadfast US opposition."[34]

The final result was a compromise achieved by the removal of any mandatory requirement to cease radar operations in the band. The final result, however, did reduce radar to a secondary status in the Table of Allocations. Primary status was restored to radar in Regions 2 and 3 for the 3.4-3.7 GHz bands, however, by footnote. The footnote included a directive for ITU Members to take all practical steps to protect the FSS after 1985 and contained a nonbinding appeal to cease operations of radar systems by the same year. Only by adding a footnote to the Table of Frequency Allocations — a method by which a country is authorized to operate in a manner other than that specified in the Table — could the United States come to terms with the various proposals made at the conference for the FSS. As part of the compromise, a formal declaration was signed by the United States, Canada, the United Kingdom, the Netherlands, Australia, and Belgium declaring that they would make reasonable effort to accommodate the FSS.[35] Thus, the status of radar was protected, but pressure from the fixed-satellite community would continue.

This dispute over the status of radar also involved the 4400-4990 MHz band. This band is extensively used by the military for tactical communications, and in NATO Europe, it has been designated as a military band. In 1963, a USSR proposal was adopted that allowed up-link fixed-satellite service use on a primary basis in the 4.4-4.7 GHz band. At the time of WARC-79, no satellite use had taken place in this band. Yet, the intentions of INTELSAT to use this band were expressed at the conference. The expressed desired use of this band for the FSS became involved with the dispute over the 3.4-3.7 GHz band. The end result was to move the FSS to the 4.5-4.8 GHz band where it would share the band with the fixed and mobile services. The band was also allocated for down-link transmissions only.[36]

The demands made by the developing countries to downgrade the status of radar systems in the 3.4-3.7 GHz band, where large investments had already been made that were vital to US national security, could be viewed as another example of the north-south tension that evolved throughout the 1970s and surfaced at the 1979 Conference. At WARC-79 the developing countries were determined to stand as a united front to effect major changes to the Radio Regulations that would ensure them their share of the radio resource for their particular needs. Not only did the developing countries band together to have great impact on the decisions related to various services in 1979, especially the fixed-satellite service, but they also set in motion the mechanisms by which a future conference could be called that would have jurisdiction over revising the regulatory procedures for space communication and the planning of certain bands for the FSS.

7.2.3 Planning of the Space Services

More than 30 resolutions and recommendations were passed at the 1979 GWARC that directly affect the space services. Besides various recommendations calling for a series of general, specialized, and regional conferences to be held in the 1980s, including a plenipotentiary conference, WARCs for the space services, conferences dealing with high frequency broadcasting and the mobile services, other regional conferences dealing with individual services, and two subregional planning conferences, two resolutions of special significance to extant space communication were passed. These were Resolution 2 and Resolution 3.

7.2.3.1 Resolution 2

Resolution 2, titled "Relating to the Equitable Uses by All Countries, with Equal Rights, of the Geostationary Satellite Orbit and of Frequency Bands for Space Radiocommunication Services," replaced the 1971 Resolution Spa 2-1 at the 1979 Conference. The 1971 Resolution provided that registration of assignments for space services should not give permanent priority to the registering country. Resolution 2, formerly titled Resolution AY, while reiterating this provision, further resolved that all countries have equal rights to use the limited, natural spectrum-orbit resource and that administrations operating communication satellite systems should "take all practicable measures to realize the possibility of the use of new space systems by other countries or groups of countries."[37]

The language of the 1979 Resolution was somewhat more strongly worded than its predecessor and gave a clear indication of the direction in which space communication policy formulation was heading. Yet, it was Resolution 3, originally titled Resolution BP, that made even a more explicit statement, indicating the direction of policy orientation for space communication — a statement that explicitly expressed the growing demand for equity in practice.

7.2.3.2 Resolution 3

Resolution 3, titled "Relating to the Use of the Geostationary-Satellite Orbit and to the Planning of Space Services Utilizing It," had its genesis amidst an extremely political climate that was characteristic of WARC-79 in general. During the 1979 Conference, a special working group of Committee 6 on Regulatory Procedures was formed "to examine all proposals related to the use of the geostationary satellite orbit and to planning of space services utilizing it."[38] The actions taken by the group of representatives from several countries that met on seven occasions during the conference and referred to as Working Group Six Ad-Hoc Two, "appears to have markedly changed the course of space communications law."[39]

7.2.3.3 Working Group Six Ad-Hoc Two

Working Group Six Ad-Hoc Two was led by T.V. Srirangan, a telecommunication expert and the Wireless Advisor of the Indian government. Srirangan was also the individual who had been nominated by the non-aligned states to be chairman of WARC-79. Despite the fact that ITU conferences had traditionally chosen their chairmen by consensus, the developing countries put Srirangan's name forward and continued to do so, even after being rejected several times. This move ended up as an unprecedented struggle over the conference chairmanship and a controversy that delayed the opening of the conference for four days.[40]

During this four-day period, a political battle occurred where the Western bloc, led by the United States, countered with a series of other candidates. One conference participant has noted that "it appeared that if the non-aligned group would not go along with a pro-Western candidate 'as a matter of principle,' they (the Western bloc) would not go along with the first of the non-aligned group's nominees." Eventually the head of the Argentinian delegation was accepted by both sides.[41] Ironically, Srirangan was thus free to be an active voice in Working Group Six Ad-Hoc Two.

In this working group, the Indian delegate as well as other influential delegates from the developing countries, put forth a proposal advocating a satellite planning conference. The force behind this proposal was initiated by dissatisfaction with the provisions of Article 9A of the Radio Regulations, which provide for the rights vesting mechanisms to the spectrum-orbit. The Indian delegation, backed by the developing countries, claimed that Article 9A did not confirm in any meaningful or realistic way the principle of equity of access to the spectrum-orbit. These countries stated that Article 9A was incapable of implementing the basic provisions for equitable access. Therefore, it was reasoned, the only practical solution was to implement a frequency-orbit allotment plan.[42]

The proposal for a satellite planning conference was limited to the fixed-satellite service, specifically the 6/4 and 14/11 GHz bands. It also included any expansion of the frequency bands for the FSS that might be decided upon at the proposed space planning conference and the planning of up-links for the broadcasting-satellite service in the 12 GHz band. The intention of the developing countries was to break away from the first-come, first-served principle of vesting rights to use the spectrum-orbit resource and to replace it with a plan that would ensure each country equal rights with all others, regardless of need, capabilities, or plans to use the resource.

In Working Group Six Ad-Hoc Two, the delegate from Colombia also called for the proposed planning conference to adopt regulations "which will guarantee rational and equitable access to the geostationary orbit for all countries . . . [and] that equitable use be made of the orbit as the Torremolinos Convention indicates; and not a pseudo-equitable use which is the case at present."[43]

Western countries countered these proposals by suggesting various methods be adopted as a solution to the problems of the developing countries. The proposal of these methods was aimed at avoiding the planning of the space services, which the developed countries did not support. These methods included the use of an elaborate multilateral coordination process and other modifications of the regulatory approach. The developing countries did not accept these suggestions.[44]

7.2.3.4 Resolution 3: Call for a Spectrum-Orbit Planning Conference

After extensive debate, the developing countries, united in their attempt to call for a planning conference, introduced a resolution for a conference in 1984 to plan the space services. The text of this resolution was produced into a final report and furnished to the Regulatory Procedures Committee of the conference. With few modifications, it was eventually adopted by WARC-79 as Resolution 3.

Resolution 3 resolved that a world administrative radio conference should be convened not later than 1984 "to *guarantee in practice* for all countries equitable access to the geostationary-satellite orbit and the frequency bands allocated to space services." The resolution called for the conference to be held in two sessions, with the first session to decide which space services and frequency bands to plan and to establish the principles, technical parameters, and criteria for the planning. Guidelines for the regulatory procedures with respect to the planned services were also to be established.[45]

Although the adoption of Resolution 3 could be viewed as part of an anticipated evolutionary process within the ITU, the significance of this resolution should not be underestimated because it, in essence, redirected the rights-vesting mechanisms for space communication. This shift of policy orientation toward *a priori* planning of the space communication services marked a change in the course of space communication law. Since the turn of the century, international recognition of rights to use the radio spectrum generally have been obtained by an *a posteriori*, case by case method, in which notification and recordation of the use of a particular frequency essentially secured the right to continued use of the registered frequency. As a result of Resolution 3, satellite communication seemed poised to join the few other services that had been planned during the attempt to engineer the spectrum, largely between 1947 and 1959.[46] (See Chapters 2 and 4.)

Table 7.1 Concerns of Preemption
(Courtesy of Stephen E. Doyle, Aerojet Tech Systems Company)

- **Systems increasing in number**
- **Perceptions of resource limits**
- **Conflicts in coordination**
- **Existing procedures unacceptable**

With the liberation of many colonial territories that became Members of the ITU occurring at about the same time as the beginning of the space age and the development of space communication, it was only a matter of time before this *a posteriori* rights-vesting method would be called into question. As the developed world exercised its technological might and began launching geostationary satellites at a rapid rate, the developing countries foresaw a time when they might be pre-empted from using the most desirable portions of the spectrum-orbit resource to meet their communication needs. At WARC-79, after more than fifteen years of expression in various international fora, the principle of equitable access was called upon to be guaranteed in practice, with the call being made for the use of an *a priori* planning conference for the space services.

7.2.4 From Technical to Political Forum

Although there are always political overtones to international conferences, until recently the ITU in particular has been characterized by the relative absence of such overtones. The very technical and physical realities inherent in the spectrum harmonization process resulted in the ITU becoming primarily a technical coordination organization, an exercise in international cooperation. Yet, WARC-79 can be said to be the ITU conference where political strategies and concerns were most widespread.

Indicative of the very political environment that characterized WARC-79 was the fact that the US delegation felt compelled to take six reservations to the decisions of the conference. This was the first time that the United States had taken such a large number of reservations. Recall that a reservation is a formal statement made by an ITU Member, indicating that the Member will not be bound to a particular decision. Thus, reservations provide a way by which countries can accept the majority of decisions reached by a WARC, while also providing an avenue that allows them not to be bound to decisions with which they cannot agree.

These reservations are called protocol statements in the Final Acts of a conference. Two of the six reservations taken by the United States at WARC-79 were directed at political issues. The other four had a direct impact on US telecommunication operations. In total, 83 reservations were included in the Final Protocol of the WARC-79 Final Acts, many of which can be characterized as political reservations.[47]

Other political aspects characterizing WARC-79 specifically involved the developing countries. As seen earlier, the new majority of developing countries had a significant impact upon the work of the ITU. In addition to the fundamental shift in majority preference from *a posteriori* to *a priori* methods used in the area of planning and managing space communication, this political impact of the third world was also measured in the political controversy over selection of the conference chairman, and in the forcefulness of spurious claims by various equatorial countries to sovereignty over the segment of the geostationary satellite orbit that lies above their respective countries. This position was forcefully maintained by a handful of equatorial countries despite the provisions in the Outer Space Treaty of 1967 and successive International Telecommunication Conventions to the effect that there can be no claim by states on Earth of sovereignty or territory over locations in space.

Nevertheless, Colombia tried to get the word "equatorial" into Resolution 3 so that it would read " . . . taking into account the relevant aspects concerning the special geographical situation of equatorial countries . . ." This move was strongly protested by the developed countries, however, and the United States demanded that the entire phrase be removed. After the USSR suggested that the word "technical" be inserted, so that the resolution would read ". . . relevant technical aspects . . . ," and that "equatorial" be replaced by "particular," the suggestion was accepted and the equatorial countries basically backed off. Consequently, the resolution was adopted to read: ". . . taking into account the relevant technical aspects concerning the special geographical situation of particular countries. . . ."[48] Nonetheless, Colombia attached two protocol statements to the resolution, stating its interpretation of it. Subsequently, the United States and many other states similarly attached statements renouncing the Colombian interpretation.[49]

Thus, the controversial issue of claimed sovereignty over the geostationary satellite orbit was deflected by political strategy. In fact, although there were a good many claims and assertions by many developing countries, in the end most did not prevail. The developing countries did not win the battle over the chairman, they did not exclude radar at lower frequencies, and no countries aside from those few equatorial states forwarding the assertion agreed to recognize sovereign claims to the GSO. Although the number of third world positions that were asserted was large, and the vehemence with which some of them were asserted was great, few such assertions were fully accepted on issues of material substance.

These controversies over the conference chairmanship and the issue of sovereignty over the GSO, however, are indicative of the shifting political orientation characterizing the ITU. As a result of this political situation, this international agency is gaining a reputation of having too much political focus and too little technical attention. If this trend toward politicization continues within the ITU, it may become ever more difficult for the Union to perform its functions and to accomplish its chartered responsibilities.

7.3 THE 1982 NAIROBI PLENIPOTENTIARY CONFERENCE

7.3.1 General Provisions Amended

Plenipotentiary Conferences do not have a technical agenda. Therefore, the 1982 ITU Plenipotentiary Conference held in Nairobi, Kenya, offered a traditional occasion for open political debate, and much political controversy did occur at the conference. One example was the argument by the Arab states for the expulsion of Israel. Although the Nairobi Conference did concern itself with such political discourse, we shall confine this discussion to the results of the conference that had direct impact on space communication.

Article 33 of the International Telecommunication Convention was amended once again in 1982, as it had been in 1973, to reflect the needs and the concerns of the developing countries with regard to equitable access to the spectrum-orbit resource. The provision in Article 33 regarding the needs and technical facilities of the developing countries that read ". . . according to their needs and the technical facilities at their disposal" was unpopular with these countries.

As a consequence, the developing countries pushed for amendment of the article on the grounds that this phrase was discriminatory. The Algerian delegate stated that the removal of any reference to needs or available technical facilities would improve or create more equal access.

The Indian delegate agreed, further stating that countries should have equal access without experiencing penalties because they lacked technical facilities at any time. Other developing countries, however, saw no reason to amend Article 33. As a result, the proposal to amend Article 33 caused strong debate at the conference, with the developed countries being generally concerned that any amendment of Article 33 that put in a phrase identifying the "special needs of the developing countries"[50] could imply inequality in favor of these countries with regard to the use of frequencies for the space services and could also imply a priority in their favor.[51]

In 1973, Article 33 of the International Telecommunication Convention read as follows: "2.) In using frequency bands for space radio services Members shall bear in mind that radio frequencies and the geostationary satellite orbit are limited natural resources, that they must be used efficiently and economically so that countries or groups of countries may have equitable access to both . . . according to their needs and the technical facilities at their disposal."[52] In 1982 the last phrase, "according to their needs and the technical facilities at their disposal," was deleted and replaced with the phrase "taking into account the special needs of the developing countries and the geographical situation of particular countries."[53]

The 1982 amendment to Article 33 of the International Telecommunication Convention could be looked at as resulting in a significant change in the meaning of the principle of equitable access. By emphasizing the special needs of the developing countries, the 1982 amendment of Article 33 required that equity be measured not simply in terms of efficiency and economy. While the principles of efficiency and economy are still retained, and are important to the general principle of equitable access, those considerations now must be viewed in light of the explicit recognition that equity specifically requires, and cannot exist without, consideration of the special needs of the developing countries.

Such an approach, however, may not always constitute the most efficient and economical use of the total global resource — if efficient and economical mean using the least amount of the spectrum-orbit resource possible, thus allowing, numerically, the greatest number of services. The use of older, less expensive technology by the developing countries may result in a somewhat less efficient use of the resource in that use of older technology necessitates the use of more of the spectrum-orbit resource than does use of the latest technology.

Also, the language "and the geographical situation of particular countries," is reminiscent of the compromise language adopted in Resolution 3 of 1979 (Resolution BP) that called for a conference for the planning of the space services. While this language cannot be interpreted as giving to the equatorial countries even slight acknowledgment of their claims of sovereignty over segments of the geostationary satellite orbit, equatorial countries would still find such language easy to support, as would other states in high latitudes or with unusual geographical shapes requiring access to a very narrow segment of the geostationary orbit, such as Chile, New Zealand, or Iceland.

Even the United States and the Soviet Union could find some solace in this language, since a very narrow segment of the GSO is available to either country that can simultaneously serve all of their respective territories, such as Hawaii, Alaska, Puerto Rico, and Maine in the case of the United States. Nevertheless, even the adoption of this geographical language is, to some extent, the result of the increased influence of the developing countries at the 1982 Plenipotentiary Conference and of their growing influence in space communication policy formulation in the ITU.

In another example, ***Article 10 of the International Telecommunication Convention*** regarding the IFRB was amended in 1982. Article 10 of the 1973 Convention read that the essential duties of the IFRB were: "a) to effect an orderly recording of frequency assignments made by the different countries so as to establish . . . the date, purpose and technical characteristics of each of these assignments, with a view to ensuring formal international recognition thereof. . . ."[54] In 1982 the reference to the date, purpose, and technical characteristics of these assignments was deleted. The amended Article 10 reads: "a) to effect an orderly recording and registration of frequency assignments made by the different countries . . . with a view to ensuring formal international recognition thereof. . . ."[55]

It is debatable whether or not the deletion of the 1973 phrase "so as to establish . . . the date, purpose and technical characteristics of each of these assignments . . ." can be interpreted as establishing a legal precedent for the diminution of the status of the *a posteriori* method of vesting rights to use the radio spectrum and the GSO. The deletion of the phrase that specified the date of an assignment, however, is clearly a strong indication that the trend toward planning that has been occurring in the ITU over the last decade is gaining momentum. The dilution of Article 10 of the International Telecommunication Convention on the IFRB is further evidence of the growing influence the developing countries are exercising, because of their numerical presence, in the area of spectrum-orbit management and in the making of space communication law.

Another area in which the developing countries had significant influence in policy making in 1982 was in technical assistance and cooperation. This influence was seen in the amending of ***Article 15*** of the Convention. Article 15 deals with the expenses of the ITU, which are met by the contributions of its Members. The 1973 Convention set the range of classes of contributory units from one-half to 30 units. The 1982 Nairobi Convention amended this contributory structure to range from one-eighth to 40 units, with the one-eighth unit class for the least developed countries as listed by the United Nations, as well as other countries determined by the Administrative Council.[56] As a result of this amendment, the developing countries were offered a greater range to choose from when selecting how much they would contribute to the finances of the ITU.

7.3.2 Resolutions, Recommendations, and Opinions

This trend toward effecting changes in ITU policy for technical assistance and cooperation was strongly expressed in the 1982 International Telecommunication Convention. Eighteen resolutions were passed specifically dealing with technical cooperation. These included various directives to the permanent organs of the ITU pertaining to the amount of time the ITU devoted to helping the developing countries with their communication needs and to the upgrading of development assistance services in the ITU.

The proposal for an ITU regional presence that had been put before the two previous plenipotentiary conferences in 1965 and in 1973 was adopted as a resolution in 1982. ***Resolution 26*** relating to an "ITU Regional Presence" resolved that "as a principle . . . a stronger presence of the Union is required in the regions to increase the efficacy of its assistance to Member countries and especially the developing ones." The Secretary-General was instructed to carry out cost-benefit and organizational studies with the objective of achieving a strengthened regional presence of the Union.[57] Thus the developing countries, garnering more political power with the passage of time, were able to get the proposal for the establishment of regional centers passed, albeit it took almost twenty years to do so.

Finally, two resolutions were adopted at the 1982 Nairobi Conference that pertained to the planning of the space services. *Resolution 1* decided that 13 future administrative radio conferences would be held before the end of the decade. Three of these were to deal with space communication. These were the Regional Administrative Radio Conference for the Planning of the Broadcasting-Satellite Service in Region 2 (RARC-83), scheduled for the summer of 1983, and the First Session of the World Administrative Radio Conference on the Use of the Geostationary-Satellite Orbit and the Planning of the Space Services Utilizing It (WARC-ORB), scheduled for the summer of 1985. The Second Session of WARC-ORB was set for the summer of 1988. Also, the next plenipotentiary conference was scheduled for 1989.

While it was decided in Resolution 1 that the First Session of WARC-ORB should formally adopt for inclusion in the Radio Regulations the decisions of the 1983 RARC on the planning of the BSS in Region 2, *Resolution 8* of the 1982 Convention instructed the Administrative Council to consider the question of planning the feeder links in Regions 1 and 3. The Council was instructed to consider the matter of the planning of the bands allocated to the FSS and reserved exclusively for feeder links for the BSS with a view to including it in the agenda of the First Session of WARC-ORB.[58]

Much of the work of the 1982 Plenipotentiary Conference was directly related to space communication. As has been shown, the level of ITU involvement in matters related to space communication paralleled the technological progress made in the field. As the space radio art advanced, relevant ITU activity increased, and, concurrently, so did the requirements and demands of the developing countries for access to the use of resources for space communication. As a consequence, the ITU has become an international forum for the mediation of these interacting factors, with the special needs of the developing countries being taken into consideration in matters related to the use of the radio spectrum and the geostationary satellite orbit for space communication.

7.4 THE 1983 REGIONAL ADMINISTRATIVE RADIO CONFERENCE FOR THE PLANNING OF THE BROADCASTING-SATELLITE SERVICE

In 1983 a Regional Administrative Radio Conference for the Planning of the Broadcasting-Satellite Service in the Frequency Band 12.2-12.7 GHz and Associated Feeder Links in the Frequency Band 17.3-17.8 in Region 2 (RARC-SAT-R2 or RARC-SAT-83) was convened. Because of opposition by the United States to the planning of DBS in this band, WARC-77 had deferred the matter of planning in Region 2 until a later conference. WARC-77, however, adopted an interim arc segmentation plan to be used in this region until a definite plan was adopted.

GWARC-79 subsequently replaced the arc segmentation plan for Region 2 with a frequency segmentation approach and expanded the total bandwidth for use in Region 2 by 500 MHz, allocating the 11.7-12.1 GHz portion to the FSS and the 12.3-12.7 GHz portion to the BSS. The 12.1-12.3 GHz segment was to be shared by both services until the 1983 RARC reached a final decision on its use.

When delegates from more than 25 countries convened in Geneva in the summer of 1983, it was expected that the five weeks of meetings aimed at planning direct broadcast satellites would be rife with debate over many issues. Using a technical report prepared by a conference preparatory meeting held by the CCIR in 1982, the RARC-SAT-83 proceeded to deal with the issues affecting use of the BSS in Region 2. Work concentrated on the technical proposals, many of which were controversial issues, with satellite signal strength the key technical issue.[59]

DBS is predicated upon the use of satellites with a higher transmitted power than satellites in the fixed-satellite service. A relationship exists between the power of the satellite and the size of the receiving earth station, leading to a trade-off between the two.

The United States approached the planning conference from the position that planning should be based on individual, direct-to-home reception, rather than on standards designed to satisfy the requirements for community reception. Therefore, the US technical proposals sought to establish at the earth's surface a satellite signal strength — or power flux density value — of minus 105 dBW (decibel watts) to the square meter. This standard would allow the use of receiving antennas of three-quarters of a meter in diameter, antennas that are less expensive than one-meter dishes. Use of this standard would permit signal strength sufficient for high-definition television (HDTV) and other multiplex analog component systems.

Canada, on the other hand, had different needs. It was concerned with setting technical parameters compatible with a community-type antenna for transmission by cable to individual homes. It was also concerned with the possible invasion of signals from the United States, if the higher signal strength were adopted. Canada, therefore, proposed a power flux density value of minus 110 dBW.[60]

Because of Canada's smaller market, expected to be about two million households *versus* the US market of up to 10 million homes, Canada opted for a lower-power DBS satellite that costs less to construct, launch, and operate, but would result in slightly higher receive costs. A compromise, called the "Brazilian compromise," was finally reached that set the signal strength value at minus 107 dBW. This two decibel difference translates into a 60 percent difference in satellite signal strength and requires a one-meter receive dish on the ground.[61]

As a result of the adoption of this lower signal strength, the United States declared it would take a reservation to the decision, that is, the US would not abide by the decision of the conference, and it would proceed with the higher power plans. This reservation was taken despite the fact that the Plan provided the US eight orbital positions, each with a maximum of 500 MHz of spectrum space that provided up to 32 channels of service.[62]

Besides setting the signal strength standard for DBS, the 1983 RARC also established the lower end of the BSS band at 12.2 GHz. This provided for a full 500 MHz of bandwidth for both the BSS and the FSS, providing for up to 32 channels of service per 500 MHz of bandwidth.[63] Thus, the provision for sharing between the two services that was established at WARC-79 was eliminated.

Greater procedural flexibility was provided by the 1983 BSS Plan for Region 2 than was provided by the 1977 Plan for Regions 1 and 3. The provisions adopted at RARC-83 were unique because although they established an *a priori* plan, it was characterized by flexibility.

The Plan contained various areas of flexibility. A system with characteristics that varied from the Plan could still be established without modification of the Plan, as long as it did not cause interference to the systems of other administrations or require more protection than a planned assignment. A system could be established that caused some interference, provided agreement with the affected administration were reached. Finally, allowance was made for some flexibility in orbital locations.[64]

Thus, greater procedural flexibility exists in the BSS Plan for Region 2 than in the Plan for the other two regions, which after ten years has not been used. A less rigid Plan in Region 2 was possible because more was understood about the need for flexibility in 1983 than in 1977. It was also possible because of a greater consensus of commitment to flexible planning in Region 2, first demonstrated in 1977 by the regional opposition to the WARC-77 Plan.

This procedural flexibility, built into the *a priori* plan for the BSS in Region 2, was important because it indicated that the definition of *a priori* planning need not be so rigid as to preclude the concept of flexibility, whereby systems that are unanticipated by, or do not adhere to, the strict characteristics of a plan may be implemented in the most efficient and economical manner. With the approach of the 1985 World Administrative Radio Conference on the Use of the Geostationary-Satellite Orbit and the Planning of the Space Services Utilizing It, such new approaches to flexible, *a priori* planning were of particular importance and would serve as a template for decisions on planning at WARC-ORB-85.

7.5 CONCLUSION

Satellite communication has become a central issue in a call for a New International Information Order, which has emerged over the last ten years at the behest of the developing countries to become a growing controversy in various international fora. Call for a new information order has been fueled by debate over information flows, specifically debate over the free *versus* the balanced flow of information across national borders.[65]

Information has been defined as a product equated with power, with differential access to information a major element in the distribution of the world's income. Many developing countries have expressed discontent with the "cultural imperialism" of the existing "order" and believe that it contributes to the homogenization of cultural values. It is argued that information should constitute a social benefit and not be treated as merchandise. This argument led to a call in the UN Educational, Scientific, and Cultural Organization (UNESCO) for a new, more just and better balanced world information and communication order with the development of advanced satellite communication playing a critical role in any movement toward such a new order.[66]

The use of satellite communication is seen as a way to equalize the flow of information. The states that call for a new information order believe that a one-way flow of information exists from the developed world to the developing world, and that the information that the developing countries receive is dictated by the developed world. This, they say, has made them subject to the cultural values of the developed world, values that are exploitive and destructive to their indigenous cultures.

Many developing countries have also recognized that satellite communication offers the potential to connect their own territories, thus allowing for the origination of information flows from within. There is a feeling that use of advanced satellite communication systems would put developing countries on a more equal economic basis with the developed world, as information is viewed as an economic asset, without which the players in any game are virtually powerless.

As a consequence of the importance of space communication in providing for even the fundamental communication requirements of the Information Age, the space services may have become an overemphasized concern of the world communications community. The entire world uses these services to maintain and to improve established communication networks; more than 25 developing countries now use INTELSAT resources to provide space services as a way to initiate the establishment of domestic, long-haul communication networks.

Space communication negotiations in the ITU, therefore, must be considered as a process — necessarily accepted by all states — in which they try to facilitate and to balance the requirements of each country. The major trend in this process since the mid-1970s has been the greatly increased influence of the developing countries, with the establishment and implementation by the ITU of policies that give special recognition to the needs of the developing world in order to help these countries catch up to the rapidly emerging technology base of the twenty-first century.

What has happened in the ITU since the mid-1970s should not be overdramatized, however. Many in the developed world believe that as time passes and facts become more clearly known, the stricter forms of *a priori* planning will be abandoned by consensus, even in the few space services where they have been employed. Such strict plans may come to be seen as counterproductive and unworkable.

Yet, significant steps toward flexible, workable planning had been taken at GWARC-79, the 1982 Nairobi Plenipotentiary, and the RARC-SAT-83. The 1979 GWARC replaced the interim arc segmentation plan in Region 2 with frequency segmentation, planned the feeder links for the BSS, greatly increased allocations for the FSS, required more precise station-keeping tolerances by satellites, and resolved in Resolution 3 to hold a planning conference for the GSO and the space services utilizing it "to guarantee in practice . . . equitable access to the geostationary-satellite orbit and the frequency bands allocated to space services."

The 1982 Plenipotentiary revised Article 33 of the Convention by "taking into account the special needs of developing countries and the geographical situations of particular countries," while dropping "according to their needs . . . and the technical facilities at their disposal." This was coupled with a revision of Article 10 of the Convention that eliminated the requirement to provide the IFRB with the date, purpose, and technical characteristics of a new national assignment. Revision of Article 15 reduced the fiscal contribution of many developing countries, and Resolution 26 stated approval of establishing regional ITU centers to assist in telecommunication development. Finally, RARC-SAT-83 adopted a flexible plan for the BSS in Region 2.

The general trend toward some form of planning for the satellite communication services, coupled with associated compromises in the ITU to advance equitable access to the spectrum-orbit, seemed to offer hope for a new consensus on satellite communication in the coming WARC-ORB and beyond. Perhaps the spirit of confrontation that was manifest at WARC-79 can be ameliorated in the future by the kinds of flexible planning characterizing the BSS Plan for Region 2 as adopted by the RARC-SAT-83.

REFERENCES

1. *International Telecommunication Convention, Montreux, 1965,* Resolution 24.
2. *Final Acts of the World Administrative Radio Conference for Space Telecommunications, Geneva, 1971,* Resolution Spa 2-1.
3. *International Telecommunication Convention, Málaga-Torremolinos, 1973,* Art. 33, (herein cited as *Málaga-Torremolinos Convention*).

4. *Radio Frequency Use and Management,* Washington, DC: Office of Technology Assessment, 1981, p. 71.
5. *Ibid.*, p. 5.
6. *Ibid.*
7. Correspondence to authors from Stephen E. Doyle, Esq., chairman of the FCC Space WARC Advisory Committee for the ITU World Administrative Radio Conference on the Use of the Geostationary-Satellite Orbit and the Planning of the Space Services Utilizing It, June 16, 1987.
8. *Final Acts of the World Administrative Radio Conference, Geneva, 1979,* (herein cited as *Final Acts WARC-79*).
9. *Radio Frequency Use and Management, supra* note 4, at 68.
10. *Ibid.*
11. *Ibid.*, p. 16.
12. See generally, "WARC '79: Curtain Going Up On Telecommunication Future, " *Broadcasting,* September 17, 1979, pp. 35-45; "Why the Sky Didn't Fall at WARC," *Broadcasting,* December 17, 1979, pp. 44-52.
13. *ITU Administrative Council Resolution 801, 1976.*
14. Codding, George A., Jr., and Anthony M. Rutkowski, *The International Telecommunication Union in a Changing World,* Dedham, MA: Artech House, 1982, p. 72.
15. *Ibid.*, p. 77.
16. *Ibid.*
17. *Ibid.*
18. Robinson, Glen O., "Regulating International Airwaves: The 1979 WARC," *Virginia Journal of International Law,* Vol. 21, 1980, pp. 1-54, at 25.
19. Johnsen, Katherine, "U.S. Seeks Better Frequency Sharing," *Aviation Week and Space Technology,* July 23, 1979, pp. 65-67, at 65.
20. *Ibid.*
21. *Ibid.*, p. 67.
22. *Final Acts WARC-79.*
23. *Radio Frequency Use and Management, supra* note 4, at 78.
24. Robinson, *supra* note 18, at 25.
25. *Ibid.*, p. 26.
26. *Ibid.*, p. 23; also *ITU Radio Regulations, 1982,* Resolutions 101 & 701.
27. *Radio Frequency Use and Management, supra* note 4, at 87.
28. *International Telecommunication Convention, Nairobi, 1982,* Resolution 8, (herein cited as *Nairobi Convention*).

29. "Efficient Use of the Geostationary Orbit," *United Nations Document A/CONF.101/BP/7,* January 16, 1981, p. 20.
30. Robinson, *supra* note 18, at 22.
31. *Radio Frequency Use and Management, supra* note 4, at 102.
32. Robinson, *supra* note 18, at 22.
33. *Ibid.*
34. *Ibid.*
35. *Radiofrequency Use and Management, supra* note 4, at 87.
36. *Ibid.*, p. 84.
37. *ITU Radio Regulations, 1982,* Resolution 2.
38. Terms of Reference for *Ad-Hoc* Group 2 of Committee 6, WARC-79 Document No. DT/150 (Rev. 1) (October 31, 1979), cited in Rutkowski, Anthony M., "Six *Ad-Hoc* Two: The Third World Speaks Its Mind," p. 1, (unpublished paper) (an edited version published in *Satellite Communications,* Vol. 4, March 1980, pp. 22-27), (herein cited as Rutkowski, "Six *Ad-Hoc* Two").
39. Rutkowski, "Six *Ad-Hoc* Two," *supra* note 38, at 1.
40. Codding and Rutkowski, *supra* note 14, at 75.
41. *Ibid.*
42. Rutkowski, "Six *Ad-Hoc* Two," *supra* note 38, at 8.
43. *Ibid.*, p. 11.
44. *Ibid.*, p. 12.
45. *ITU Radio Regulations, 1982,* Resolution 3 (emphasis added).
46. Various other services that have been planned include the Aeronautical, High Frequency Broadcasting, Maritime-Mobile, and Radiotelegraph Services.
47. *Radiofrequency Use and Management, supra* note 4, at 69.
48. Rutkowski, Six *Ad-Hoc* Two," *supra* note 38, at 31.
49. Robinson, *supra* note 18, at 16.
50. *International Telecommunication Convention, Nairobi, 1982,* Art. 33, (herein cited as *Nairobi Convention*).
51. Smith, Milton J., "Space WARC 1985: The Quest for Equitable Access," *Boston University International Law Journal,* Vol. 3, 1985, pp. 229-255, at 40.
52. *Málaga-Torremolinos Convention,* Art. 33.
53. *Nairobi Convention,* Art. 33.
54. *Málaga-Torremolinos Convention,* Art. 10.
55. *Nairobi Convention,* Art. 10.
56. *Ibid.*, Art. 15.
57. *Ibid.*, Resolution 26.
58. *Ibid.*, Resolution 8.

59. "RARC 83: Cooperation Eliminates Surprises," *Satellite Communications,* July 1983, pp. 22-27, at 24.
60. "US Runs Into Roadblocks Over DBS Plans," *Broadcasting,* July 4, 1983, p. 23, at 23.
61. *Ibid.*
62. *Final Acts of the World Administrative Radio Conference on the Use of the Geostationary-Satellite Orbit and the Planning of the Space Services Utilizing It,* Appendix 30.
63. *Ibid.*
64. *Ibid.*
65. See generally, Masmoudi, Mustapha, "The New World Information Order," *World Communications A Handbook,* (G. Gerbner & M. Siefert, eds.), New York: Longman, 1984, pp. 14-27.
66. "The New World Order for Information," Tunisian Ministry of Information, *United Nations Document A/SPC/33/L.5,* October 20, 1978.

Chapter 8

The 1985 Space Conference — To "Guarantee In Practice" Equitable Access to Space Communication

8.1 INTRODUCTION

Two developing countries, India and Indonesia, were the prime impetus behind the call of WARC-79 for an orbital planning conference. India and Indonesia held serious concerns, based on experience, about their future ability to use the spectrum-orbit resource for their communication needs. Indonesia was beset by protracted negotiations with Intelsat concerning placement of its domestic-regional Palapa system. Likewise, when India set out to launch its first domestic communication satellite, it had problems finding a satisfactory orbital position as the most preferable slots were already taken by the USSR and by INTELSAT. After inharmonious coordination negotiations, India not only wound up with a slot it considered inferior for its very specialized satellite, which contained a meteorological, remote sensing, and communication platform, but also India wound up agreeing to reduce its proposed bandwidth and satellite power.[1] It was little wonder that India led the call at WARC-79 for guaranteed equitable access to the geostationary satellite orbit.

India, Indonesia, and other developing countries expressed concerns about the rapid increase of systems in the GSO, perceptions of resource limitation, and conflicts in coordination. Concern was also expressed that regulatory procedures in place were unable to provide for equitable access to the spectrum-orbit resource and were therefore unacceptable.[2]

All these problems were laid at the door of the ITU in the hope that they could be solved by this historically effective, technical coordinating body. The ITU was the natural forum for the debate that resulted, a debate primarily triggered by intransigent behavior of the USSR and by the complex technical coordination required with a global system like INTELSAT. The real issues were the Intelsat space segment guard band and spacing requirements and the unresponsiveness of the Soviet Union, which refused to comply with the ITU recommendations to adjust to accommodate latecomers.[3] Had ITU regulations and recommendations been implemented, particularly by the USSR, the 1979 problem might not have been nearly as bad as it was, and there might not have been a call for a WARC-ORB.

Nevertheless, since WARC-79, a great increase has occurred in the number of geostationary communication satellites and in the consequent request for orbital slots. Sixty percent of some 150 communication satellites in the GSO have been launched since 1979.[4] From 1980 to 1985, the number of requests for satellite orbital slots for C and Ku-band satellites more than doubled. While many of these proposed satellites may never be launched or even constructed, at the end of 1984, 90 percent of some 200 proposed satellites belonged to the developed countries or were controlled by international organizations, with 50 percent of these belonging to just nine countries.[5] This situation only increased the resolve of the developing countries to achieve some form of planning at the orbital planning conference.

The World Administrative Radio Conference on the Use of the Geostationary-Satellite Orbit and the Planning of Space Services Utilizing It (WARC-ORB) was to be held in two sessions. The first session would take place in 1985, with the second session scheduled for 1988. The changes in the regulatory regime for space communication that were ultimately agreed upon at WARC-ORB-85 were negotiated over a period of three years through a CCIR special working party, a Conference Preparatory Meeting in 1984 and the First Session of the Conference in 1985.

8.2 THE 1985 WORLD ADMINISTRATIVE RADIO CONFERENCE ON THE USE OF THE GEOSTATIONARY-SATELLITE ORBIT AND THE PLANNING OF THE SPACE SERVICES UTILIZING IT

The First Session of the WARC-ORB was held in Geneva from 8 August to 15 September 1985. The Second Session was scheduled for mid-1988.[6] More than 800 delegates assembled in Geneva for WARC-ORB-85 to carry out the primary responsibilities of the conference.

WARC-ORB-85 has been characterized as the most important space telecommunication conference of the century.[7] It has been seen in this light because the purpose of this conference was to decide the direction that space communication regulation would follow probably well into the next century.

The negotiations at this conference took place in an environment in which two diametrically opposed positions were required to seek accommodation. On the one hand, developing countries called for *a priori* planning for the frequency bands and geostationary orbital positions for space communication to ensure equitable access to both. On the other hand, several countries, including the United States, held that such planning, if not carefully limited, could inhibit technological progress by forcing technology to meet artificial and unnecessary constraints rather than allowing it to proceed in an orderly manner while protecting the need for access to the spectrum-orbit. It was as a result of these opposing positions that a compromise, hybrid regulatory approach for space communication was recommended by the First Session, WARC-ORB-85, for final consideration of the Second Session, WARC-ORB-88.[8]

While some of the developing countries persisted at WARC-ORB-85 in further establishing their demand for planning of the spectrum-orbit resource, the substantial majority of countries seemed willing to limit the nature of the planning approach, making it more flexible and less *a priori* in nature, and to limit the extent of this planning to newly allocated FSS expansion bands. Thus, the regulation of space communication embarks upon a new era, an era that is characterized by compromise and change aimed at balancing the requirements of the developed and the developing countries in a rational, effective, regulatory regime based upon consensus.

8.2.1 Agenda of the Conference

The agenda of the conference, as set by the Administrative Council and approved by ITU Members, was strongly related to achieving a guarantee of equitable access to the spectrum-orbit resource for space communication. The agenda provisions were taken almost directly from Resolution No. 3 adopted at WARC-79, which called for a conference to plan the space services. The primary responsibilities of the First Session of WARC-ORB were:

- to decide which space services and frequency bands were to be planned;
- to establish the principles, technical parameters and criteria for the planning and to provide guidelines for associated regulatory procedures;
- to establish guidelines for regulatory procedures for the space services and frequency bands not identified for planning; and
- to consider other possible approaches that could meet the objective of guaranteed equitable access to the spectrum-orbit resource.[9]

After deciding which space services and frequency bands to plan, the conference was to select a planning method and to establish the principles, technical parameters, criteria, and guidelines for associated regulatory procedures. With respect to the bands and services selected for planning, the conference also had to specify the form (technical parameters, date of bringing assignment into use, *et cetera*) in which requirements of administrations should be submitted and the date of submission, with the date for submission allowing sufficient time for an administration to formulate its requirements. With reference to the services and frequency bands that were not identified for planning, the conference had to establish guidelines for regulatory procedures "as necessary."[10]

8.2.2 Preparation for the Conference

A complex mix of issues had to be considered in preparing for WARC-ORB-85. In consideration of these issues, several factors played an essential role. Gained experience and technological advances in satellite communication offered valuable insight for preconference activities. Many administrations had come to feel that regulatory flexibility was necessary to accommodate the rapid change brought about by these advances, although they were willing to accept that some form of limited planning might be appropriate in selected areas. Many administrations continued to view accommodation as generally preferable to planning, believing that nongeostationary satellite services did not require any planning whatsoever.

Because of all the issues to be considered, as the activities leading up to the 1985 Space WARC began, several specialized organizations lent their help and technological expertise in preparing for the conference. As part of the preconference activities, a CCIR Conference Preparatory Meeting (CPM) was convened in Geneva in the summer of 1984 to produce a report that could serve as a technical basis for WARC-ORB-85. The CPM, a special joint meeting of nine CCIR Study Groups, was held from 25 June to 20 July 1984. It was attended by more than 300 delegates from 61 administrations and from 33 recognized private operating agencies and international organizations.[11]

Although the CCIR was directed to carry out studies to provide technical information for the 1985 Conference, it should be noted that the IFRB was also expressly given the responsibility to prepare a report for the conference "on the operation of the procedures of Articles 11 and 13 including information about difficulties which may be reported to the IFRB by administrations in gaining access to suitable orbital locations and frequencies, and to circulate this report to administrations at least one year before the First Session of the Conference."[12]

This supplemental role of the IFRB to provide for the technical basis of WARC-ORB primarily came about at the impetus of the developing countries, which felt that the CCIR could create a bias in the outcome against planning by introducing technical reasons why planning was not feasible. Recall that because of financial and technical ability, it is the developed nations that most frequently attend CCIR meetings. It is these same nations that view strict or rigid planning as stifling to technological innovation. As a result the IFRB, which aids the developing nations in frequency selection and registration, was chosen to carry out the survey of experience to date and identification of problems for presentation to WARC-ORB-85.[13] This IFRB report was considered an important balance for the potentially technical preparation work of the CCIR.

The Conference Preparatory Meeting of the CCIR in 1984 produced seven proposals for space communication regulation that were considered at WARC-ORB-85. These proposals related to alternative planning approaches. Five of these proposals were developed by a CCIR working party in 1983-84, with the other two proposals coming, one each, from the USSR and China. The CCIR proposals were characterized by the following specific elements:

1. a long-term, 10-20 year *a priori* allotment plan;
2. a short-term, 3-5 year *a priori* allotment plan;
3. a detailed *a priori* allotment plan with spectrum-orbit segmentation;
4. guaranteed access by means of multilateral coordination; and
5. phased in revision as necessary of existing regulatory procedures, regulations and CCIR recommendations.

The USSR proposal combined *a priori* planning with the possibility of flexible system design, providing for the revision of technical parameters and regulatory procedures to accommodate new requirements. Finally the Chinese proposed a worldwide plan that would cover one satellite generation lifetime of about ten years.[14]

8.2.3 Conference Results

General agreement was finally reached at WARC-ORB-85 on a hybrid approach to planning. The planning method adopted consisted of three parts and embraced two novel approaches to spectrum management — a frequency allotment plan to be drawn up only for the FSS in the expansion bands newly allocated for this service at WARC-79 and a multilateral planning method to be used as a means of improving and simplifying procedures for coordination requirements in other designated FSS bands.[15]

WARC-ORB-85 decided to focus for planning on the fixed-satellite service only. Although at the conference many developing nations proposed *a priori* planning of the 6/4, 8/7, 14/11-12, and 30/20 GHz bands for the FSS, the conference ultimately adopted a more flexible arc allotment plan limited to the expansion bands allocated for the FSS in 1979, namely 4500-4800 MHz, 10.70-10.95 GHz, and 11.20-11.45 GHz for down-links and 6425-7075 MHz and 12.75-13.25 GHz for up-links.[16]

Recall that an *a priori* plan assigns specific orbital slots and radio frequencies for use by individual administrations, thus specifically and rigidly prescribing the use of the spectrum-orbit resource. On the other hand, the arc allotment plan adopted by the 1985 Space WARC provided for more flexibility, alloting instead of prescribed orbital slots, the use of a predetermined arc segment of several potential orbital slots and frequency bands.

The arc allotment plan offered substantially more flexibility in the use of the spectrum-orbit resource than did strict *a priori* planning by allowing allotments to be made from a range of frequencies within a predetermined arc. This move from rigid, *a priori* planning toward flexible arc allotment planning that is only quasi *a priori* in nature resulted as a compromise at WARC-85 aimed at satisfying the resource requirements of the developing countries as well as accommodating the needs of all countries for enough flexibility to enable technological innovation to proceed without severe constraints and allowing for adjustments to changing needs over time.

Thus, the outcome of WARC-ORB-85, as enunciated in its Report to the Second Session of the Conference, combined two ideologically opposed positions to arrive at a potentially acceptable accommodation for all countries. Once again, the nature of radiocommunication had intervened to encourage mediation. Regulation of radiocommunication cannot be successfully conducted as a political contest, with winners and losers and untenable compromises. The use and management of the spectrum-orbit resource is a zero-sum game, the outcome of which is of vital interest to every country in the world. Once again, at WARC-85, this reality encouraged international cooperation. Selected developing countries achieved their goal of establishing some form of planning for space communication, while other countries, including many developing countries, succeeded in making sure that the planning method adopted was not strictly *a priori* in nature, but provided for sufficient flexibility in spectrum management.[17]

Also, the United States and allied countries succeeded in keeping the 3.4-3.7 and the 8/7 GHz bands from being planned, as was proposed by some developing countries. It is in these bands that the United States and a number of other countries carry on military communication services, such as radar communication. Therefore, the US position was that it was critical to its national security that this portion of the spectrum not be subjected to planning. Further, the conference deferred the possible planning of the 30/20 GHz band for the FSS, calling for more study on the matter.[18]

8.2.3.1 The Arc Allotment Plan

One part of the planning method suggested to the Second Session by WARC-ORB-85 was an arc allotment plan. This plan would provide each administration with at least one orbital position within a predetermined arc and a frequency bandwidth of at least 800 MHz within predetermined bands so that administrations could satisfy their communication requirements for national service. The allotment plan was established in the following expansion bands that were allocated to the FSS at WARC-79:

- 4500 - 4800 MHz (down-links);
- 6425 - 7075 MHz (300 MHz to be selected from this band for up-links);
- 10.70 - 10.95 GHz (down-links);
- 11.20 - 11.45 GHz (down-links); and
- 12.75 - 13.25 GHz (up-links)[19]

It should be noted, however, that the United States and other developed countries opposed including the 14/11-12 GHz expansion band in the bands designated for allotment planning because they felt that the 6/4 GHz expansion band made sufficient spectrum space available for planning. As a consequence, the United States filed a declaration, reserving the position that it will take with respect to this band at the second session of the conference. France and the United Kingdom took a similiar position, stating that they would wait to see how much flexibility was built into the allotment plan submitted by the second session.[20]

These declarations were only notices of concern, yet they could result in a reservation taken to the Final Acts of the Second Session. A reservation is a notification that a country does not consider itself bound by the conference decision with which it disagrees. Reciprocally, when such a reservation is taken, the reserving Member may not seek or obtain enforcement of the provisions reserved.

Provision for a guarantee of access was ensured by the allotment planning method accepted at WARC-ORB-85. There was, however, insufficient participation for a valid vote to be taken on this provision as it was adopted toward the end of the conference when many delegations had already left to return home. The conference was planned to conclude on

Friday, 13 September 1985, but protracted plenary sessions extended day and night through Saturday and Sunday, concluding finally at 2:30 A.M. on Monday, 16 September.[21] Many delegates had travel arrangements in place for Saturday and Sunday departures from Geneva. Thus, a lack of a quorum resulted during the final plenary sessions. Nonetheless, the Report to the Second Session of the Conference does specify that quarantee of access be an integral element of the planning method.

Although no quorum was attained on this provision, the implications of this procedural defect are probably not so great for the second session of the conference, because the document forwarded to the second session exists as a report, not a binding international treaty. Thus, the findings in the Report to the Second Session of the Conference will serve as the basis for the international space communication law that is agreed upon in 1988 and carries great weight. Where the lack of a quorum may have impact is in allowing for the issue regarding guarantee of access to arise again. Although an alternative proposition was maintained by two administrations, it is, however, more than likely that the provision will remain as written, as it is backed by the majority of developing countries, sufficient in number to reach a quorum.

To guarantee equitable access to the GSO, all ITU Members were given at least one allotment in the proposed plan consisting of three specific elements:

1. an orbital position within a predetermined arc;[22]
2. a bandwidth of 800 MHz;[23] and
3. a service area limited to national systems providing domestic services.[24]

Reference in the plan to a predetermined arc was made as a means of increasing the flexibility of the plan. Specification of the size and position of the arc was left to intersessional study that was to take place between the first and the second sessions of WARC-ORB.[25]

The IFRB was directed by the First Session of WARC-ORB to carry on planning exercises during the intersessional period. With respect to the establishment of a predetermined arc, the planning method was to examine the possibility of establishing orbital arcs associated with the orbital positions necessary to cover the service area of individual administrations, while being compatible with other allotments in the plan and allowing for subregional systems.[26]

The allotment plan specifically limited the service area to national systems providing domestic services. Provisions were also to be made in the procedures of the plan permitting administrations with adjacent territories to combine all or part of their allotments so as to ensure a subregional service.[27] The IFRB was thus directed to carry out planning exercises for an allotment plan that would provide for each country, or geographical area, one coverage with one or more beams. Administrations were to submit to the IFRB individual descriptions of the requirements of their coverage area.[28]

When carrying out the planning exercise, the IFRB was also directed to attempt to provide the maximum bandwidth of 800 MHz to each allotment. If this was not possible to achieve, however, identical bandwidths were to be provided to each allotment.[29] Thus, the allotted bandwidth could ultimately result in less than 800 MHz.

The arc allotment plan adopted at WARC-ORB-85 specified that the duration of the plan was to be for a period of at least ten years. However, there was insufficient participation toward the end of the conference to reach a quorum on this matter when it was put to a vote. Ultimately, it was left to the second session of the conference to decide on the exact duration of the plan. It was also recommended that the second session of the conference adopt provisions to ensure that unless specifically agreed to by an administration concerned, any allotment to satisfy its national requirements would not be deleted.[30]

The Report to the Second Session of the Conference stipulated that the overall aim of the improved procedures part of the allotment plan shall be to "guarantee in practice for all countries equitable access to the orbit/ spectrum resources in the relevant bands, at the same time protecting existing systems."[31] One way to protect existing systems that was recommended by the conference was to adopt provisions to ensure that any allotment to satisfy an administration's national requirements should not be deleted without the agreement of the administration concerned.[32]

This recommendation, however, requires careful thought for several reasons: it could result in a *de facto* sovereignty, or permanent priority, to a portion of the spectrum-orbit resource; the UN Outer Space Treaty of 1967 rejects claims of territory and sovereignty over locations in space by nations on Earth;[33] WARC-ST-71 Resolution Spa 2-1 and GWARC-79 Resolution 2 clearly indicate that registration of use of the spectrum-orbit should not and does not confer any permanent priority over any particular portion of the resource; and the Report to the Second Session of WARC-ORB also indicated that existing systems " . . . shall be included in the plan on an equal basis with planned allotment and may be subject to some adjustments. . . ."[34]

8.2.3.2 Regulatory Procedures

The regulatory procedures associated with the planning method were also left to be established at the second session of the conference. WARC-ORB-85, however, did stipulate that such procedures should include the following:

- procedures to be followed by administrations that wanted to modify their allotments appearing in the plan;
- procedures to be followed for converting an allotment into an assignment;
- procedures to be followed in order to ensure that new Members obtain an allotment in the plan; and
- procedures applicable to additional FSS users in the bands covered by the allotment plan, that is, users of FSS services other than those for domestic services.[35]

Already existing systems in the bands established for the allotment plan that were recorded in the Master Register or for which the coordination procedure had commenced or advance publication had been received by the IFRB before 8 August 1985 were to be included and given equal status with planned allotments. Although, technically, existing systems would be protected, any system might nonetheless be subject to some adjustments, with the degree of adjustment depending upon the stage of its development. The specific adjustment criteria were left to be drawn up by the second session of the conference.[36]

It was stipulated in the Report to the Second Session of the Conference that the planning method must preserve the rights of other services sharing equal, primary status in the bands to which the allotment planning method was to be applied. To accommodate this provision, it was foreseen that appropriate sharing criteria would need to be adopted and applied.[37]

Thus, although the actual allotment plan would be worked out at the second session of WARC-ORB, the first session of the conference was successful in specifying the frequency bands in which the allotment plan was to be established and in enunciating associated parameters to be considered when developing the plan. Every country was to be alloted a right of access to at least one orbital position in the GSO within a predetermined arc, with the possibility of obtaining more than one allotment depending upon an administration's requirements for one coverage of its geographical area.[38] Provision was also made for the establishment of procedures to ensure that countries that were not Members of the ITU at the time the plan was drawn up could obtain allotments.[39]

The planning method proposed would ensure that when any country wanted to establish a national satellite network in the FSS within the period of validity of the plan, it would have access to sufficient spectrum-orbit resource to do so. To implement such a system, an administration would publish details of the intended network parameters, including the exact orbital location desired, and negotiate with the administrations sharing its predetermined arc or with those with allotments in adjacent arcs. After reaching agreement, the details of the assignment would be notified to the

IFRB for registration in the Register. It was foreseen that the sharing by administrations of the various orbital arcs would be designed to allow administrations to establish cooperatively, within the framework of the allotment plan, subregional satellite networks serving several countries.

This provision is particularly relevant to countries intending to establish their first satellite network, as it would allow administrations to combine their allotments. In this way a country that was not yet able to use the total payload capacity of a single satellite could join its allotment with those of other countries to create a subregional system that could provide services to all cooperating administrations, thus allowing for a country's first national satellite network to be set up.

8.2.3.3 Planning by Improved Procedures

The other part of the planning method recommended by WARC-ORB-85 consists of improved coordination procedures to be applied to the bands allocated to the FSS in the 6/4 and 14/11-12 GHz bands not subject to the allotment plan. These are the conventional bands other than the expansion bands. In these bands, improved coordination procedures are proposed to allow administrations to satisfy their communication requirements if they need more bandwidth than is provided for by the allotment plan. The procedures would be applied in bands that are already in general use for FSS networks. They are:

(C band)
- 3700 - 4200 MHz,
- 5850 - 6425 MHz;

(Ku band)
- 10.95 - 11.20 GHz,
- 11.45 - 11.70 GHz,
- 11.70 - 12.20 GHz in Region 2,
- 12.50 - 12.75 GHz in Regions 1 and 3,
- 14.00 - 14.50 GHz;

(Ka Band)
- 18.10 - 18.30 GHz,
- 18.30 - 20.20 GHz,
- 27.00 - 30.00 GHz.

The conference also decided to call for further study by the CCIR of the technical character of the FSS in the 30/20 GHz band and required the CCIR to report its findings to the second session of the conference with the view to reaching a decision on the future planning of these bands. It was therefore left to the second session of WARC-ORB ultimately to decide whether or not to subject this Ka band allocated to the FSS to planning or simply to improved coordination procedures.[40]

A principal characteristic of proposed improved procedures is the holding of periodic multilateral planning meetings to coordinate new satellite systems on a multilateral rather than bilateral basis. The multilateral planning method, proposed by the United States and a number of other developed nations, was seen as a way to ease procedural and political complications of bilateral coordination.

The Report to the Second Session of the Conference recommended that the multilateral planning meeting ". . . shall be the normal process for gaining access to the GSO-spectrum resources." The Report also recommended that the multilateral planning meeting approach " . . . should be a new and separate procedure to be added to the Radio Regulations." The procedural aspects of the meetings and the status of their decisions were left to the second session of the conference to decide.[41]

The adoption of multilateral coordination procedures would result in an entirely new approach to the traditional bilateral coordinating mechanisms used in the ITU for over eighty years. Multilateral coordination is a procedure based on coordination of new satellite systems in a group as opposed to the current case-by-case, bilateral coordination of new systems. The multilateral planning method would place all the satellite systems in the group — for instance all those proposed in a given time period, in a given service, for operation in the same bands — on an equal basis rather than on a first-come, first-recognized basis, where earlier systems have an advantage over later systems.

The overall aim of the improved procedures planning method is to guarantee in practice for all countries equitable access to the spectrum-orbit resource in the frequency bands designated at WARC-ORB-85 for planning while protecting existing systems. Certain guidelines for the formulation of future regulatory procedures were established at the first session to ensure the aim of guaranteed access. Some of the possible features suggested in the Report to the Second Session to be included in these regulatory procedures were:

- simplification of the advance publication procedure;
- consideration of multilateral planning meetings;
- burden sharing between administrations to share the inconvenience of the adjustments required to enable the accommodation of new systems;
- consideration of existing systems in the designated bands;
- use of technical measures to resolve problems of coordination; and

- provisions for bilateral coordination between administrations with the availability of assistance by the IFRB.[42]

It has been foreseen that the impact of the improved procedures planning approach, where multilateral planning meetings are used to gain access to the spectrum-orbit resource, could be tremendous. Therefore, the Report to the Second Session urges administrations to consider the matter further during the intersessional period and to submit proposals regarding multilateral planning meetings to the second session. The proposals should consider various factors relevant to such meetings, such as the timing of the meetings, the status of the decisions taken, the financial implications, the scope and form of administrations' requirements and the stage at which they should be submitted, the participants in the meeting, the safeguarding of interests of nonparticipants, the convening authority, and whether bilateral agreements reached in the period between meetings should be subject to ratification at the next meeting.[43]

8.2.3.4 Planning Principles

Both of the planning methods adopted at WARC-ORB-85 — the arc allotment plan and planning by improved procedures — are to be based upon certain principles adopted by the conference. The planning principles enunciated at the 1985 Conference were:

- guarantee of access and equitability;
- sharing with other services;
- reservation of resources;
- the technical aspects of special geographical situations;
- consideration of existing systems;
- provisions for multiadministration systems;
- flexibility;
- different planning solutions in different circumstances;
- efficiency;
- provisions for multiservice and multiband networks; and
- administrative cost as low as possible.[44]

Elaboration upon these principles, as set forth in Chapter 3 of the Report to the Second Session of the Conference, follows. With respect to the principle of **guaranteed access and equitability**, it was stipulated that the planning method shall guarantee in practice for all countries equitable access to the geostationary satellite orbit and the frequency bands allocated to the space services utilizing it. The special needs of developing countries and the geographical situation of particular countries are also to be taken into account.

The principle of **sharing with other services** means that where frequency bands allocated to one space service using the geostationary satellite orbit are also allocated to other space services or to terrestrial services on an equal, primary basis, the planning methods must fully respect the equality of rights to operate in these bands. The planning method and associated regulations, therefore, must not impose additional constraints on services sharing bands on an equal basis.

Reservation of resources means that the planning method must be consistent with the universally accepted principle that administrations or groups of administrations are not entitled to permanent priority in the use of particular frequencies and GSO positions in such a way as to foreclose access by other administrations to the resources. Therefore, the planning method should consider the full spectrum-orbit resource, with the possibility of setting aside portions of the resource for unforeseen requirements or for requirements of future Members only to be considered after all current requirements are met.

The principle of **technical aspects of special geographical situations** refers to the consideration of relevant technical aspects of the geographical situation of particular countries, like those in high latitudes where only specific orbital positions can be used to service the area. As noted in Section 3.4.2.1.3 of the Report to the Second Session, countries in special geographical situations may find that economies realized in the manner of planning and use of the space segment may be no greater than economies realized on the ground. The Conference Preparatory Meeting Report to WARC-ORB-85, at Annex 4, Section 4.5, elaborates the implications of these considerations. Special geographical situations of importance could include:

- special latitudes;
- dispersed territories;
- terrain obstructions;
- precipitation and sandstorms;
- geographically small countries;
- geographically large countries;
- elongated countries; and
- dispersed population centers.

When the substantial majority of the participants contemplated the "special geographical situations" of countries, it was with these kinds of factors and their economic and operational aspects in mind. The equatorial countries, promoting the issue of their declared sovereignty over superjacent arcs of the GSO above their countries, essentially ignored the factors listed above, when they claimed that the agreed reference to "special geographic situations" was an acknowledgment of their claims. In plain fact, the factors relating to permanent priority that were discussed above, such as the Outer Space Treaty, Resolution Spa 2-1 of WARC-ST-71, and Resolution 2 of WARC-79, when combined with the action of the WARC-ORB-85 Plenary Session in denying the Conference's competence to address the issue of sovereignty, clearly exclude the equatorial countries' claims. During the Plenary, the ITU Secretary-General provided a legal opinion that the Conference did not have competence in the matter of sovereignty because it was a legal, as opposed to technical, matter, that was not on the agenda of the conference. The 1985 Conference thus decided by consensus to notify the UN that it had no competence over the issue of sovereignty.[45]

As noted above both in Section 8.2.3.1 and also in Section 8.2.3.2, the 1985 Conference also decided that the planning principles should include **consideration of existing systems**. Such consideration should not exclude the possibility that, if necessary, existing systems may be subject to some adjustments to accommodate new systems. A series of factors relevant to such adjustments is presented and discussed in the Report to the Second Session,[46] which notes that "these factors can give benefit only if all or most satellite networks operating in a frequency band support them; the burden must be shared."[47]

The principle of **flexibility** means that the planning method should provide the means to accommodate unforeseen requirements and the modification of requirements of administrations. The planning method should also be capable of accommodating advances in technology, while not preventing the use of well proven and widely used technologies.

The use of **different planning solutions in different circumstances** means that while a worldwide planning solution would be best, the possibility of using different planning methods for different regions, frequency bands, or orbital arcs should not be excluded. **Efficiency** means that the planning method should ensure efficient and economical use of the spectrum-orbit resource allocated to the space services, and **provisions for multiservice and multiband networks** refers to the accommodation of multiservice or multiband satellite networks, without imposing undue constraints on planning.

Finally, the principle of **provisions for multiadministration systems** refers to the consideration of requirements of administrations using multiadministration systems created by intergovernmental agreement, such as

INTELSAT and EUTELSAT (a European organization for international satellite communication). This principle provides that such systems are to be used collectively without affecting the rights of administrations with respect to their national systems. The planning method must also take into account the specific characteristics of multiadministration systems to enable them to continue to meet the requirements of administrations for international services as well as for national services. Such multiadministration systems include those having a safety-of-life aspect, like INMARSAT, and those having feeder links in the FSS.

8.2.3.5 Provisions for Multiadministration Systems and the Issue of International Systems Separate from INTELSAT

The provisions incorporated into the planning principles for multiadministration systems resulted from a compromise over proposals from different administrations. A Swiss proposal, backed by some 30 delegations, wanted common user-type organizations like INTELSAT that are essential to small countries with no other means of access to communication satellite service, to be clearly recognized to ensure that they were included in the planning process. These 30 delegations proposed that such multiadministration satellite organizations be given special recognition with respect to present and future requirements for geostationary orbital positions and radio frequencies necessary to provide international service. They also proposed that the spectrum-orbit resources necessary for multiadministration organizations to provide domestic services be treated equally with the requirements of administrations planning to operate their own domestic systems.[48]

The United States felt that this proposal afforded preferential treatment to multiadministration systems, thus establishing a dangerous precedent. Support for the Swiss proposal at WARC-ORB-85, however, was strong. Therefore, the United States sought a compromise so as to avoid taking a reservation on the matter. In the course of events, the United States succeeded in adding the phrase "without affecting the rights of administrations with respect to national systems" into the planning principle provisions for multiadministration systems to avoid the establishment of an undesirable precedent favoring any single class of spectrum-orbit user. Although the term "national systems" is not defined, "[t]he United States and others repeatedly stated that they used the term 'national systems' to include <u>all</u> systems that a nation might choose to establish." The US position was to ensure the protection of domestic and private international systems such as INTELSAT, INMARSAT, and others.[49] Thus, the United States did not interpret the phrase "national systems" to mean exclusively domestic systems.

The debate over the treatment of multiadministration satellite systems was seen privately by some members of the US delegation to WARC-ORB-85 as being politically inspired — raised as a way to block new, private international satellite systems separate from INTELSAT that had been proposed by US entities. These separate systems would provide international services in competition with multiadministration organizations, particularly INTELSAT.[50] Perhaps the reason for this perception of some delegates was the contemporary international outcry that had arisen over the US decision to, for the first time, permit private international satellite systems separate from INTELSAT.[51]

At the time of the interim creation of INTELSAT in 1964, and during and after the negotiation (1968-71) and entry into force (1973) of the permanent, definitive INTELSAT Agreement and Operating Agreement,[52] the United States consistently maintained the position that the INTELSAT agreements contemplated a single, global, commercial communication satellite system. INTELSAT was envisaged as something of a beneficent monopoly, the sole entity for communication via satellite between nations. The view was espoused that other international systems were unnecessary and at odds with the creation of INTELSAT. The justification for this claim of exclusivity is found in INTELSAT's scheme of participatory membership and the representation of member countries in the governance of the organization, as well as in INTELSAT's promise of nondiscriminatory pricing and nondiscriminatory, worldwide access.[53]

This participatory nature of INTELSAT membership and governance is seen in its structure, which consists of an Assembly of Parties, a Meeting of Signatories, a Board of Governors, and an Executive Organ. The Assembly of Parties is composed of member countries and meets every two years to consider long-term and constitutional issues. The Meeting of Signatories is composed of the designated telecommunication entities of member countries, such as COMSAT in the United States, and meets once a year to consider operational, technical, and financial issues. The Board of Governors in 1987 is an executive committee of 27 representatives from member countries that meets four times a year to make decisions concerning design, development, establishment, operation, and maintenance of the global INTELSAT space segment. The Executive Organ, a kind of Secretariat with an internationally recruited staff, is headquartered in Washington, D.C. It is headed by a Director General, who is the chief executive and legal representative of the INTELSAT organization.[54]

As of 1987, the INTELSAT organization was comprised of 112 countries, all of which are part owners. A minimum share (0.05%) of INTELSAT cost about US $750,000 in 1985.[55] Such a payment represents a capital contribution that is proportional to an ownership share in the satellite cooperative. With certain exceptions, the amount of capital investment is adjusted annually to reflect the percentage of use made of the system by each respective member. Utilization charges from various services are in addition to ownership costs and constitute INTELSAT's operating revenues.[56]

INTELSAT received preferred treatment or protected status in its own multinational Agreement, which is an international telecommunication treaty, even before the efforts at WARC-ORB-85 to extend that preference to the Radio Regulations of the ITU. Article XIV(d) of INTELSAT's Agreement required:

> To the extent that any Party or Signatory or person within the jurisdiction of a Party intends individually or jointly to establish, acquire or utilize space segment facilities separate from the INTELSAT space segment facilities to meet its international public telecommunications services requirements, such Party or Signatory, prior to the establishment, acquisition or utilization of such facilities, shall . . . ensure technical compatibility of such facilities and their operation with the use of the radio frequency spectrum and orbital space by the existing or planned INTELSAT space segment and to avoid significant economic harm to the global system of INTELSAT. . . .[57]

Two-thirds approval of the Assembly of Parties, the plenipotentiary organ of INTELSAT, is required for a formal *recommendation* of approval of the use of a separate international satellite system by a Party to the INTELSAT Agreement. By the time of the first tentative and limited US approval of separate systems in 1985, several satellite systems separate from INTELSAT, primarily regional in nature, had already been approved by the INTELSAT Assembly of Parties.[58]

This protected status of INTELSAT had been promoted by the United States — the prime architect of the INTELSAT Agreements — for many years prior to 1985. INTELSAT was established at the impetus of the United States in 1964 as a "single global commercial communications satellite system" to "provide expanded telecommunications services to all areas of the world, . . . contribute to world peace and understanding . . . [and] provide, through the most advanced technology available, for the benefit of all nations of the world, the most efficient and economical service possible consistent with the best and most equitable use of the radio spectrum. . . ."[59]

The principles of economy and equity were uniquely embodied in the final INTELSAT Agreement, which requires nondiscriminatory access to the INTELSAT system. Article V of the INTELSAT Agreement provides as follows:

> The rates of space segment utilization charge for each type of utilization shall be the same for all applicants for space segment capacity for that type of utilization.[60]

This provision requires that the same rates apply to the same type of use *and that they be the same for all users of that type of use*. The practice of INTELSAT has been to equate types of use with different types of service or service offering, and it has been generally held that Article V is a provision for the benefit of the developing world. The US Communications Satellite Act of 1962, which was the first step in the establishment of INTELSAT, clearly directed that "care and attention will be directed toward providing such services to economically less developed countries and areas . . ."[61]

Therefore, the separate systems issue at WARC-ORB-85 was more than a political issue raised by some 30 administrations. Indeed, the implementation of private separate satellite systems that would compete with multiadministration INTELSAT-type organizations in providing international telecommunication services was considered controversial in that implementation of such systems would be a reversal of long-standing US policy. This was seen by some as a direct threat to developing countries that benefit from nondiscriminatory, indirectly subsidized INTELSAT rates. Since, by charter, INTELSAT cannot discriminate in the rates that it charges, all customers are charged the same rates, regardless of distance or traffic volume. An indirect subsidy results because heavily used routes are not permitted a discount under this system. The result is a theoretical subsidy for the thin, or lightly used routes, such as those to remote, sparsely populated areas in the developing world.

There was some concern at WARC-ORB-85, especially among those countries proposing a preference for multiadministration systems, that if separate systems were implemented in the United States, then INTELSAT might be forced to discount its rates on the heavy volume traffic routes to compete with such separate systems. Even though a change in the INTELSAT charter can be effected by a two-thirds vote of the Assembly of Parties, fears arose that the necessity for competition might cause an amendment to the pricing provisions of Article V, the consequent removal of the subsidy to thin traffic routes, and increased rates to those routes, if the heavy route rates are discounted.

Indeed, despite the majority of developing countries in the Assembly of Parties of INTELSAT and the difficulties this majority could pose toward any attempt to amend the INTELSAT Agreement, the US Congress passed an act, signed by the President, that instructed the Secretary of

State to consult with COMSAT and the Secretary of Commerce "regarding the appropriate scope and character of a modification to article V(d) of the INTELSAT Agreement that would permit INTELSAT to establish cost-based rates for individual traffic routes, as exceptional circumstances warrant, paying particular attention to the need for avoiding significant economic harm to the global system of INTELSAT as well as to United States national and foreign policy interests."[62]

The rapid growth in satellite applications and the growth of regional systems around the world led the United States to consider limited competition to INTELSAT in the provision of international satellite services. Despite the seeming contradiction of the new US position with respect to its traditional insistence on preferential treatment for INTELSAT, the fact is that, as far back as 1962, the US Communications Satellite Act had contemplated that the United States and other countries would establish separate domestic satellite systems and also might someday establish other international satellite systems. The act stated that it did not "preclude the creation of additional communications satellite systems, if required to meet unique governmental needs or if otherwise required in the national interest."[63]

The President of the United States determined on 28 November 1984 that "separate international communications satellite systems are required in the national interest."[64] Following this determination, the US Federal Communications Commission (FCC), after consultation with the Secretary of State, authorized US separate systems in principle, but placed significant limits on such systems to comply with US treaty obligations under Article XIV(d) of the 1971 Intelsat Agreement. US separate systems were to offer only services that were unavailable or uneconomical through INTELSAT, were forbidden to connect with public switched networks (public telephone networks), and were required to obtain the prior permission of correspondent countries before a license would be granted by the FCC.[65] Thus, as it became clear that separate US systems would be limited, few, and likely to have little influence on the global rate structure of INTELSAT, it became easier to achieve compromise language on the consideration given in planning multiadministration systems.

Again, the provisions finally incorporated into the planning principles provided that "the planning method shall take into account the requirements of administrations using multi-administration systems created by inter-governmental agreement and used collectively without affecting the rights of administrations with respect to national systems." The provisions also provided that the specific characteristics of multiadministration systems be considered "to enable them to continue to meet the requirements of administrations for international services as well as, in many cases, for national services."[66]

8.2.3.6 Incorporation into the Radio Regulations of the 1983 Direct Broadcast Satellite Plan for Region 2

The direct broadcast satellite Plan for the western hemisphere adopted at the 1983 Regional Administrative Radio Conference was approved and incorporated into the Radio Regulations at WARC-ORB-85. Approval and incorporation of the Plan into the Regulations, however, was not achieved without controversy. Several European nations, notably the United Kingdom and France, feared that satellite systems operating under the 1983 Region 2 Plan would cause interference to their DBS, FSS, and terrestrial services. Only after study and discussion of the potential interservice sharing problems, solutions were found, and the Plan was approved.

With the incorporation of the 1983 Region 2 DBS Plan into the Radio Regulations, broadcasting-satellite service assignments in the western hemisphere were included in the Master Register of the IFRB and protected against interference from systems subsequently authorized. Additionally, with the incorporation of the 1983 Plan into the Regulations, the precise constraints necessary to protect systems in Region 1 (Europe, Africa, the Middle East, the USSR, and Mongolia) and in Region 3 (Asia and the Pacific) against interference were set forth.[67]

8.2.3.7 Planning of the Feeder Links for the 12 GHz Broadcasting-Satellite Service in Regions 1 and 3

Another issue related to the broadcasting-satellite service was on the agenda for WARC-ORB-85. As mandated, the first session of the conference was to define the most suitable technical characteristics for the feeder links, or up-links, to broadcasting satellites in Regions 1 and 3. The conference also was to identify those bands for which sharing criteria between services needed to be developed during the intersessional period for consideration at the second session of the conference.[68]

WARC-ORB-85 decided to select the 14.5-14.8 and 17.3-18.1 GHz frequency bands for the feeder link assignment plan for Regions 1 and 3. Malta and countries outside Europe were to use the 14.5-14.8 GHz band. The conference decided not to use the 10.7-11.7 GHz band made available at WARC-79 for the feeder link assignment plan.[69]

The 1985 Space Conference recommended that administrations use the 17.3-18.1 GHz band as far as possible, as the 300 MHz available in the 14.5-14.8 GHz band did not have enough capacity to feed all the channels assigned per country in the plan, if a large number of countries chose to use this band. The conference also suggested that exclusive use of the 17.3-18.1 GHz band for feeder links — a band that was also selected for DBS feeder links in the 1983 Plan for Region 2 — left more operating latitude for terrestrial services sharing the band on a primary basis with the FSS. Thus, the second session of WARC-ORB was directed by the 1985 session to start BSS feeder link planning by using the 17.3-18.1 GHz band.[70]

8.2.3.8 Recommendation Plen/B Relating to High Definition Television in the Broadcasting-Satellite Service

At the 1971 World Administrative Radio Conference for Space Telecommunications, the 22.5-23 GHz band was allocated to the broadcasting-satellite service in Region 3 in addition to the 12 GHz band that was allocated to all three regions. The 1977 12 GHz DBS Plan for Regions 1 and 3 was based upon a 625-line picture standard for television. Since 1977, the 22 GHz band has been looked upon in Region 3 as a band that could provide for high definition television (HDTV), which needs a much wider channel bandwidth than that necessary for the 1987 CCIR 625-line standard. The western hemisphere also looks to the 22 GHz band as a potential HDTV band.

Although not explicitly on its agenda, WARC-ORB-85 took up the matter of HDTV. The conference recognized that interest in HDTV by satellite was rising in Europe and that techniques for it were rapidly progressing. The conference also noted that the frequency bands presently allocated to the BSS do not as planned provide for a worldwide allocation suitable for the implementation of a worldwide standard for high definition TV transmission by satellite.

Therefore, in *Recommendation Plen/B*, WARC-ORB-85 called for the consideration of making a frequency allocation around 22 GHz available worldwide for the BSS to accommodate high definition television and recommended that the Administrative Council place the matter on the agenda of the second session of the conference. Recommendation Plen/B also called on the CCIR to report to the second session of WARC-ORB on the development of technical parameters for HDTV transmissions by satellite and on the frequency bands appropriate for its use, as well as on aspects of inter and intraservice sharing.[71]

8.2.3.9 Recommendation Plen/C Relating to Satellite Sound Broadcasting Systems for Individual Reception by Portable and Automobile Receivers

At WARC-79, Resolution 505 was adopted concerning satellite sound broadcasting systems for individual reception by portable and automobile receivers. This resolution encouraged administrations to carry out experiments with a broadcasting-satellite sound service within the 500-2000 MHz band. The 1979 Resolution also invited the CCIR to continue its studies on the technical characteristics of a satellite sound broadcasting system for individual reception and on the feasibility of sharing with terrestrial services, as well as on appropriate sharing criteria. WARC-ORB-85 was authorized to consider the results of the various studies on the subject and to make any decisions regarding the allocation of a suitable frequency band for a broadcasting satellite sound service.[72]

Satellite sound broadcasting has been looked at as a potentially more economical alternative to terrestrial sound broadcasting. Studies have shown that the concept is feasible within the optimum frequency range of 500-2000 MHz. Implementation of such a service that would share a frequency band with other services, however, would present extensive problems of interference because of the powerful down-link emission that would be necessary. To accommodate a broadcasting-satellite sound service within this frequency range, sufficient width within the band would have to be cleared. So far sufficient bandwidth within the optimum frequency range has not been found.

In consideration of the matter of sound broadcasting from satellite to individual portable receivers, WARC-ORB-85 adopted *Recommendation Plen/C.* This recommendation stated that while satellite sound broadcasting is technically feasible, at present the allocation of an exclusive band to a broadcasting-satellite sound service on a worldwide basis was not possible. Therefore, the search for a suitable frequency band should continue. Administrations were advised to continue their studies on the matter and to consider the use of frequencies within and also outside but near the 500-2000 MHz range. The second session of the conference was left to consider the results of the various studies and to make appropriate decisions on the matter of satellite sound broadcasting to individual portable and automobile receivers.[73]

8.2.3.10 Recommendation Plen/A: Draft Agenda for the Second Session of the Conference

As a result of the extensive amount of work dealt with by WARC-ORB-85, of necessity, most of the decisions related to this work were left to the second session of the conference. *Recommendation Plen/A*, a draft agenda for WARC-ORB-88, was adopted by the 1985 Space WARC to serve as the basis of the work of the second session. The 1988 Space Conference now faces the tremendous responsibility of accomplishing its agenda.

In carrying out its responsibilities, WARC-ORB-88 must consider the following items:

- proposals from administrations;
- the Report of the first session of the conference;
- preparatory work carried out in the intersessional period;
- the relevant reports from the IFRB and the CCIR; and
- the requirements for the allotment plan submitted by administrations.[74]

These items, however, must be considered in light of the situation that some of the frequency bands subject to planning are allocated on a shared basis with equal rights to more than one space service, and most are also allocated with equal rights to terrestrial services. These rights will need to be taken into account when devising any plan.

The draft agenda for the second session of the conference consists of several items, including:

1. the establishment of an allotment plan and associated regulatory procedures for the FSS in the frequency bands identified for planning at WARC-ORB-85;
2. the establishment of improved regulatory procedures for the FSS in the frequency bands identified at the first session of the conference;
3. the adoption of appropriate technical standards, parameters, and criteria pertaining to the FSS in those bands identified for planning;
4. the review and revision as necessary of regulatory procedures and technical standards, parameters, and criteria pertaining to space services and frequency bands not subject to planning;
5. the review and revision as necessary of definitions relating to space services;
6. the establishment of provisions and an associated plan for the feeder links in Regions 1 and 3 and the incorporation of the plan into the Radio Regulations, as well as the consideration of any appropriate amendment of the Radio Regulations related to the plan;
7. the consideration of studies and the making of any appropriate decisions related to sound broadcasting in the broadcasting-satellite service;
8. the consideration of and any appropriate action on the subject of a suitable worldwide frequency band in the broadcasting-satellite service to accommodate high definition television in the 22 GHz band;
9. the amendment of the Radio Regulations as necessary to incorporate the decisions of the second session of the conference; and
10. the evaluation of the financial impact of the decisions of WARC-ORB-88 on the budget of the ITU.[75]

8.2.4 Intersessional Studies

The second session of WARC-ORB was charged with the tremendous responsibility of accomplishing a great deal of work in a very short time. The ultimate success of WARC-ORB-88 will depend to a great extent upon the technical work that is conducted during the intersessional period by the IFRB and the CCIR, as well as on the administrative tasks performed by the Administrative Council and the General Secretariat.

Not only must the Administrative Council establish the final agenda, meeting dates, and budget for WARC-ORB-88, but it must also supervise the intersessional activities of the IFRB and the CCIR and supply the financial backing for this work. The General Secretariat, in addition to providing all the necessary administrative support for the intersessional activities, must also coordinate the increased use of ITU computer facilities and publish and distribute all the reports compiled during the intersessional period.

Although the responsibilities of these two administrative organs of the ITU are necessary to the ultimate success of the second session of the conference, the major responsibility for the success of WARC-ORB-88, however, was charged to the IFRB and to the CCIR. These two organs of the Union face an enormous task.

8.2.4.1 Intersessional Duties of the International Frequency Registration Board

The IFRB was given several major tasks to perform during the intersessional period. It was charged with developing a software program for the preparation of the allotment plan and with carrying out appropriate planning exercises. In view of the limitations of manpower, time, and financial resources, the IFRB staff could only produce a developed, tested program on a timely basis if Member administrations assisted and contributed resources to these ends. During 1986 and 1987 representatives from Japan, France, Switzerland, and the United States were actively supporting the IFRB staff in these efforts.

To accomplish its task, the IFRB must convene various meetings to describe the computer programs to be developed and the schedule of work to be done, to review the progress of such work, and to discuss initial tests of the computer programs to determine their suitability for the second session of the conference. It also must meet to conduct a final review of the intersessional program. In addition to these meetings to be convened to consider intersessional activities, the IFRB must also establish the technical parameters upon which the planning effort is to be based.[76]

The IFRB as well must perform a range of duties relevant to the success of establishing a plan for the space services. These duties include reviewing the sharing criteria for the space and terrestrial services,[77] aiding in the development of guidelines for regulatory procedures associated with the planning method,[78] and performing several important duties associated with the planning of the broadcasting-satellite feeder links in Regions 1 and 3, such as adapting computer software for feeder link planning exercises and performing the exercises.[79]

In addition, the IFRB has been charged with reviewing the Master Frequency Register, the International Frequency List, and other related documents to make sure that such registries accurately reflect the current use of the spectrum-orbit resource. The IFRB thus must assess, with reference to the space services, which assignments are not in use in order to cancel or modify them with the agreement of the administration concerned. An accurate, up-to-date record is necessary to enable the second session to carry out its work.[80]

8.2.4.2 Intersessional Duties of the International Radio Consultative Committee

The work to be carried out by the CCIR during the intersessional period is just as extensive as that of the IFRB. Besides those studies cited earlier, such as the study of the technical character of the FSS in the 30/20 GHz band, the CCIR must conduct more than a dozen technical studies related to the establishment of a successful allotment plan.

In addition, the CCIR must look into interservice sharing criteria,[81] conduct various studies related to devising a feeder link plan for the 12 GHz broadcasting-satellite service in Regions 1 and 3,[82] undertake studies to define the system parameters for satellite sound broadcasting,[83] and carry out studies relevant to the development of technical parameters for high definition satellite television.[84] The extent of these tasks is immense, and the CCIR will need to call upon all of its resources in order to accomplish the work.

8.3 CONCLUSION

The extent of the work to be carried out by the organs of the ITU during the intersessional period of the WARC-ORB is unprecedented, requiring extensive organization, coordination, technical expertise, and financial backing. In addition to the tremendous amount of work to be done, the most difficult obstacles facing the ITU organs in carrying out the suggestions of WARC-ORB-85 are the short time period between sessions and the question of who is going to pay for all the intersessional activites. As it is now, the ITU is pressed to stay within its budget. The added expense of these interim studies could significantly exceed the financial resources of the ITU.

Although ITU Members will most likely be asked to take on additional expenses, the United States, as one of the heaviest contributors to the ITU, has already expressed its position on the matter. The stated US policy has been to hold budgets of United Nations agencies to increases only at the level of inflation. Selected agency and private industry experts, however, have contributed programming, planning, and other expert talent to support the IFRB and the CCIR in their intersessional work. The possibility exists that other countries will take similiar approaches.

WARC-ORB-85 was a conference where both developed and developing countries came together from extreme positions to arrive at a compromise in order to effect changes to the regulatory regime for space communication. The conference was only partially successful, however, in achieving its agenda goals. Like GWARC-79, WARC-ORB-85 was beset by political problems. Much of the technical work assigned to Committee 5 of the Conference was not concluded, because the Committee was not able to complete its work due to inharmonious proceedings. Once again, political problems proved more troublesome than technical problems.

The general feeling held by the developed world after WARC-79 was that the ITU, once largely a technical forum free of politics, is becoming increasingly politicized as communications become more important to governments. This feeling was not abated by WARC-85. Many in the developed world continued to feel that the political quotient of ITU conferences has increased with the number of ITU Members, the majority of which are developing countries. Indeed, the head of the US delegation to WARC-ORB-85 noted that the United States with its size and population would be treated the same under the allotment plan adopted by the conference as Malta. The ITU, he observed, "is one country, one vote. It is concerned with equality, not equity."[85]

Indeed, whether or not the ITU in its present form continues as a successful exercise in international cooperation may very well depend upon the results of WARC-ORB-88 and associated proceedings. There was talk among the US delegates in private during WARC-ORB-85 concerning the possible failure of the conference and the possibility of, and the need for, fundamental restructuring of the ITU.[86] Failure of the first session was avoided, however, by compromise born of a realization among developing countries that failure would be more damaging to them than to the developed countries. The developing countries realized that failure of the conference would simply leave in place the existing system of allocating slots and frequencies on a first-come, first-served basis, a system that the developing world has found fault with more and more in recent years.

Given this realization, and with plans for the BSS already adopted, it was readily agreed at the first session of WARC-ORB in 1985 that only the FSS appeared to require any form of planning, and that within the FSS, only the 1979 expansion bands would be planned for the present. The FSS is regulated by the same frequency coordination procedures that are used for most of the other space radio services. These procedures are set out in Articles 11 and 13 and associated Appendices 3, 4, 28, and 29 of the ITU Radio Regulations of 1982.

One exception was made to this general regulatory approach of limiting the plan to the FSS expansion bands. The exception applies to the feeder links that feed program signals to broadcasting satellites. Although these feeder links use frequencies that are allocated to the FSS, an *a priori* Region 2 satellite broadcasting frequency assignment plan for the 12 GHz band that included a plan for the feeder links in the 17.3-17.8 GHz band was drawn up at RARC-83, approved at WARC-ORB-85, and incorporated into the Radio Regulations.[87]

At WARC-ORB-85, various changes to the regulatory regime for the fixed-satellite space communication service were agreed to. (See Table 8.1.) The conference agreed that, in the future, several frequency bands allocated to the FSS should be regulated according to a new regime that embraces three different methods — a quasi-*a priori* frequency allotment method of planning involving allocation of arc segments rather than of specific orbital locations, improved coordination procedures to ensure equitable access possibly involving provisions for new multilateral planning meetings, and the continued use of the traditional coordination procedures. The suggested arc allotment plan would include a guarantee of access for each country to the spectrum-orbit with:

- an orbital position within a predetermined arc;
- a bandwidth of up to 800 MHz; and
- a service area limited to national systems providing domestic services.[88]

Therefore, despite the unfulfilled objectives, political overtones, and daunting intersessional tasks, the WARC-ORB-85 represents a giant step toward rational, consensus-based, international arrangements for space communication that will guarantee in practice equitable access to the GSO and the space services utilizing it. If the WARC-ORB-88 is equally successful, a milestone should have been reached in an evolutionary process from principle to practice that began even before the first Extraordinary Administrative Radio Conference on Space Communication in 1963.

Table 8.1 Key Results of WARC-ORB-85
(Courtesy of Stephen E. Doyle, Aerojet Tech Systems Company)

- **Planning in the Fixed Satellite Service only**
- **Arc Allotment Plan - Not A Priori Plan**
- **Arc Allotment Plan for expansion bands only**
  -4.5 to 4.8 GHz (down) **-300 MHz in 6.425 to 7.075 GHz band (up)**
- **Improved planning procedures in:**

3.700 - 4.200 GHz	**11.45 - 11.70 GHz**
5.850 - 6.425 GHz	**12.50 - 12.75 GHz**
10.95 - 11.20 GHz	**14.00 - 14.50 GHz**

- **Further study to consider possible planning in 20 to 30 GHz band**
- **No planning activity in 3.4 to 3.7 GHz or in the 8/7 GHz bands.**
- **Incorporation of RARC-BS Region 2 Final Acts into Radio Regs.**

REFERENCES

1. *First Report of the Advisory Committee for the ITU's World Administrative Radio Conference on the Use of the Geostationary-Satellite Orbit and the Planning of the Space Services Utilizing It,* December 1983, pp. 4-20 to 4-25. (herein cited as *First Advisory Report*).
2. *Radiofrequency Use and Management,* Washington, DC: Office of Technology Assessment, 1981, p. 71.
3. *First Advisory Report, supra* note 1, at 4-23 to 4-24.
4. Demac, Donna A., George A. Codding Jr., Heather E. Hudson, and Ram S. Jakhu, *Access to Orbit: After the 1985 ITU Space WARC,* London: International Institute of Communications, 1986, p. 8.
5. *Report of the United States Delegation to the First Session of the ITU World Administrative Radio Conference on the Planning of the Geostationary Satellite Orbit and the Space Services Utilizing It*, 1986, p. 23, (herein cited as *US Delegation Report*).
6. World Administrative Radio Conference on the Use of the Geostationary-Satellite Orbit and the Planning of Space Services Utilizing It, First session, Geneva, 1985, *Report to the Second Session of the Conference*, (herein cited as *Report to the Second Session*).
7. "Satellite Slots," *Fortune,* June 24, 1985, p. 88.
8. *Report to the Second Session, supra* note 6, Chap. 3.3.
9. *ITU Administrative Council Resolution 895, 1983.*

10. *Ibid.*
11. *Report of the CCIR Conference Preparatory Meeting (CPM), Joint Meeting, Study Groups 1,2,4,5,7,8,9,10 and 11,* Geneva, 25 June - 20 July 1984, p. 2, (herein cited as *CCIR CPM Report*).
12. *ITU Administrative Council Resolution 895, 1983.*
13. Robinson, Glen O., "Regulating International Airwaves: The 1979 WARC," *Virginia Journal of International Law,* Vol. 21, 1980, pp. 1-54, at 37.
14. *CCIR CPM Report, supra* note 11, at 37-40.
15. *Report to the Second Session, supra* note 6, Chap. 2, p. 3; Chap. 3.3.
16. *Ibid.*
17. See *Report to the Second Session, supra* note 6, Chap. 3; and Rutkowski, Anthony M., "Six *Ad-Hoc* Two: The Third World Speaks Its Mind," *Satellite Communications,* March 1980, p. 22.
18. *US Delegation Report, supra* note 5, at 59.
19. *Report to the Second Session, supra* note 6, Chap. 3.3.
20. "WARC 1985: The Politics of Space," *Broadcasting,* September 23, 1985, pp. 56-58, at 57.
21. Correspondence to the authors from Stephen E. Doyle, Esq., chairman of the FCC Space WARC Advisory Committee for the ITU World Administrative Radio Conference on the Use of the Geostationary-Satellite Orbit and the Planning of the Space Services Utilizing It, June 16, 1987.
22. *Report to the Second Session, supra* note 6, Chap. 3.3.1(a).
23. *Ibid.*, Chap. 3.3.4.4.
24. *Ibid.*, Chap. 3.3.4.1.
25. *Ibid.*, Chap. 3.3.4.5.
26. *Ibid.*, Chap. 8 (2.5).
27. *Ibid.*, Chap. 3.3.4.1.
28. *Ibid.*, Chap. 8 (2.1).
29. *Ibid.*, Chap. 8 (2.4).
30. *Ibid.*, Chap. 3.3.4.6.
31. *Ibid.*, Chap. 3.3.5.2.
32. *Ibid.*, Chap. 3.3.4.6.
33. Treaty on Principles Governing the Activities of States in the Exploration and Use of Outer Space, Including the Moon and Other Celestial Bodies, 18 UST 2410, TIAS No. 6347, 610 UNTS 205 No. 8843 (signed at Washington, London, and Moscow, January 27, 1967), Art. 2.
34. *Report to the Second Session, supra* note 6, Chap. 3.3.4.9.
35. *Ibid.*, Addendum, Annex 1.
36. *Ibid.*, Chap. 3.3.4.9.
37. *Ibid.*, Chap. 3.3.3.

38. *Ibid.*, Chap. 8 (2.1).
39. *Ibid.*, Chap. 3.3.4.7.
40. *Ibid.*, Chap. 3.3.1.(b).
41. *Ibid.*, Chap. 3.3.5.1.
42. *Ibid.*, Chap. 3.3.5.3.
43. *Ibid.*, Chap. 3.3.5.4.
44. *Ibid.*, Chap. 3.2.1 - 3.2.11.
45. *US Delegation Report, supra* note 5, at 91.
46. *Report to the Second Session, supra* note 6, Chap. 3.4.2.1.1.
47. *Ibid.*, Chap. 3.4.2.1.2.
48. Demac, Codding, Hudson and Jakhu, *supra* note 4, at 10.
49. *US Delegation Report, supra* note 5, at 56.
50. "The Way It Was and Wasn't at WARC '85," *Broadcasting,* November 4, 1985, pp. 70-72, at 71.
51. *Report and Order in the Matter of Establishment of Satellite Systems Providing International Communications,* CC Docket No. 84-1299, FCC 85-399, September 3, 1985, (herein cited as *FCC Report and Order on Separate Systems*).
52. *Agreement Relating to the International Telecommunications Satellite Organization "Intelsat"* (Intelsat Agreement), *August 20, 1971,* (entered into force *February 12, 1973)* 23 UST 3813, TIAS No. 7532, and *Operating Agreement Relating to the International Telecommunications Satellite Organization "Intelsat"* (Intelsat Operating Agreement), *August 20, 1971,* (entered into force *February 12, 1973)* 23 UST 3892, TIAS No. 7582, (herein cited as *Intelsat Agreement* and *Intelsat Operating Agreement*).
53. Colino, Richard R., "A Chronicle of Policy and Procedure: The Formulation of the Reagan Administration Policy on International Satellite Telecommunications," *Journal of Space Law,* Vol. 13, 1985, pp. 103-156, at 104.
54. *Intelsat Report 1984-1985,* p. 20.
55. *Ibid.*, pp. 32-36.
56. *Intelsat Operating Agreement, supra* note 52.
57. *Intelsat Agreement,* Art. XIV(d).
58. For a complete list of satellite coordinations completed under Art. XIV(d) of the *Intelsat Agreement* from 1973 through 1983, see *Colino, supra* note 53, at 105 note 10.
59. *Agreement Establishing Interim Arrangements for a Global Commercial Satellite System* (Interim Agreement), signed *August 20, 1964,* 15 UST 1705, TIAS No. 5646, (herein cited as *Intelsat Interim Agreement*).
60. *Intelsat Agreement,* Art. V(d).

61. Pub. L. No. 87-624, 76 Stat. 419, 1962, (codified at 47 USC Sections 701-704) (herein cited as *Comsat Act*).
62. *Foreign Relations Authorization Act, Fiscal Years 1986 and 1987*, Pub. L. No. 99-93, Sec. 146, 99 Stat. 405, 1985.
63. *Comsat Act, supra* note 61, Sec. 102(d).
64. *Presidential Determination, No. 85-2, November 28, 1984.*
65. *FCC Report and Order on Separate Systems, supra* note 51.
66. *Report to the Second Session, supra* note 6, Chap. 3.2.6.
67. *Final Acts of the World Administrative Radio Conference on the Use of the Geostationary-Satellite Orbit and the Planning of Space Services Utilizing It, Geneva, 1985,* (herein cited as *Final Acts WARC-ORB-85*).
68. *ITU Administrative Council Resolution 895, 1983.*
69. *Report to the Second Session, supra* note 6, Chap. 6.1.3.1.
70. *Ibid.*, Chaps. 6.1.3.3.1.1; 6.1.3.3.1.3; and 6.1.3.3.2.1.
71. *Ibid.*, Recommendation Plen/B.
72. *Final Acts of the World Administrative Radio Conference, Geneva, 1979,* Resolution 505.
73. *Report to the Second Session, supra* note 6, Recommendation Plen/C.
74. *Ibid.*, Recommendation Plen/A.
75. *Ibid.*
76. *Ibid.*, Chap. 8.
77. *Ibid.*, Chap. 4.
78. *Ibid.*, Annex 1.
79. *Ibid.*, Chap. 6.
80. *Ibid.*, Resolution Com5/1.
81. *Ibid.*, Chap. 5.
82. *Ibid.*, Chap. 6.
83. *Ibid.*, Chap. 7 and Recommendation Plen/C.
84. *Ibid.*, Recommendation Plen/B.
85. "WARC 1985: The Politics of Space," *supra* note 20, at 57.
86. "The Way It Was and Wasn't at WARC '85," *supra* note 50, at 71.
87. *Final Acts WARC-ORB-85, supra* note 67.
88. *Report to the Second Session, supra* note 6, Chap. 3.3.4.3.

PART IV
SPECIAL SPACE COMMUNICATION TOPICS

Chapter 9

Direct Broadcast Satellites and the Evolution in the United Nations of International Space Law

9.1 INTRODUCTION

The launch of Sputnik on 4 October 1957 sent a shock wave reverberating through the international system. Discussion began almost immediately within the United Nations on the subject of formulating legal principles to be applied to outer space. Within a year, the first outer space resolution met with consensus in the General Assembly.

On 13 December 1958 General Assembly Resolution 1348 (XIII) was adopted. The resolution emphasized that outer space was the common interest of mankind and noted that it should be used only for peaceful purposes and for the benefit of all mankind. The resolution recognized that outer space activites could increase knowledge and improve life and it encouraged the pursuit of cooperative international and scientific programs.[1] The resolution also created an 18-member United Nations ad hoc Committee on the Peaceful Uses of Outer Space.

Attitudes and concepts were expressed in this first outer space resolution that have, to this day, been reflected in much of the work of the United Nations and that serve as guidelines for the entire development of space law. Of specific import was:

- the awareness that man's existence exceeded that of his terrestrial birth;
- that there existed the potential for increased practical benefits in this new dimension;
- that international cooperation was necessary to realize this potential; and
- that the intention of providing a guiding framework of legal principles for the exploration of outer space was to avoid "the extension of the present rivalries in this new field." (G.A. Resolution 1348)[2]

9.2 THE COMMITTEE ON THE PEACEFUL USES OF OUTER SPACE

The coordinative committee of the General Assembly of the United Nations that deals with all aspects of space related issues is the Committee on the Peaceful Uses of Outer Space (COPUOS). With beginnings in 1958 as an ad hoc committee, COPUOS was officially made a permanent committee of the United Nations when General Assembly Resolution 1472 (XIV) was passed on 12 December 1959. In the text of the resolution it was stated that space was to be used for "the betterment of mankind" and for "the development of science and the improvement of the well-being of peoples."[3]

Two permanent subcommittees constitute COPUOS: the Legal Subcommittee and the Scientific and Technical Subcommittee. COPUOS is a public international committee made up of official representatives from more than 50 nation-states.

It is to be noted that COPUOS was established in the United Nations at about the time that space activities (the launch of Sputnik) and the space race between the United States and the Soviet Union began. Within this context, the workings of COPUOS evolved to be highly political in nature, especially the work of the Legal Subcommittee.

The early years of COPUOS were marked by continual disputes over membership and voting criteria, as well as by controversies over the substantive issues of international policy. Nevertheless, the Legal Subcommittee of the ad hoc COPUOS recommended in 1959 that:

1. the United Nations Charter and the Statute of the International Court of Justice not be confined to Earth;
2. extensive study of the principles and procedures that apply to the sea and to airspace be conducted to determine their relevance to space regulation; and
3. the initial creation of a comprehensive code of space law is impractical, but a set of general principles could be developed in response to the practical problems arising in this new environment.[4]

Yet, the enunciation of general principles was stalled by an impasse concerning the decision-making procedure to be followed. The Soviet Union insisted on an absolute veto within COPUOS for all states and refused to participate in any substantive proceedings until such a procedure could be agreed upon. The United States, on the other hand, maintained that some form of majority voting procedure should be used, even if the threshold for decision ultimately should be other than a simple majority, such as two-thirds or three-fourths.

In 1961, after several years of disagreement on this issue between the two space powers, a compromise accord was finally reached. It was agreed that, contrary to normal practice in the United Nations, the work of COPUOS would flow according to the principle of consensus.[5]

According to this principle, no member had a legal right of veto; yet, in practice, no recommendation should go forward to the General Assembly if any member strenuously objected. Therefore, all the recommendations of COPUOS would be essentially unanimous and so carry great weight. Although the principle of consensus agreement is not a formal approach, the members of COPUOS decided that consensus would be more effective than decisions made by majority vote, as the contentious issue of formal veto power would be avoided. COPUOS has operated according to this principle ever since.

Sputnik was also the impetus for early suggestions that satellites would some day carry broadcast signals directly to home receivers in foreign countries (DBS — Direct Broadcast Satellites). Writers in the USSR wrote in regard to this future technology:

> With the help of a large Sputnik . . . Moscow television programs could easily be relayed not alone to any point in the Soviet Union, but also far beyond its borders.[6]

The issue of direct broadcast satellites was thus taken up in the United Nations within the general context of the organization's involvement in outer space affairs and the development of space law.

9.3 THE UNITED NATIONS AND THE INTERNATIONAL TELECOMMUNICATION UNION

In addition to COPUOS, a standing committee of the UN General Assembly, the other major source of international law applicable to space communication law and DBS, as we have seen, flows from the charter of the International Telecommunication Union (ITU Convention) and its Radio Regulations. As previously noted also, the International Telecommunication Union is a public international organization and a specialized agency of the United Nations. All specialized agencies of the UN are autonomous, treaty-based, public international organizations that are affiliated with the UN through an instrument of agreement with the UN Economic and Social Council. (See Figure 9.1.) The ITU has worked closely with the Legal Subcommittee of the UN-COPUOS in the consideration of policy formulation, regulations and treaty provisions that bind ITU's Member governments.

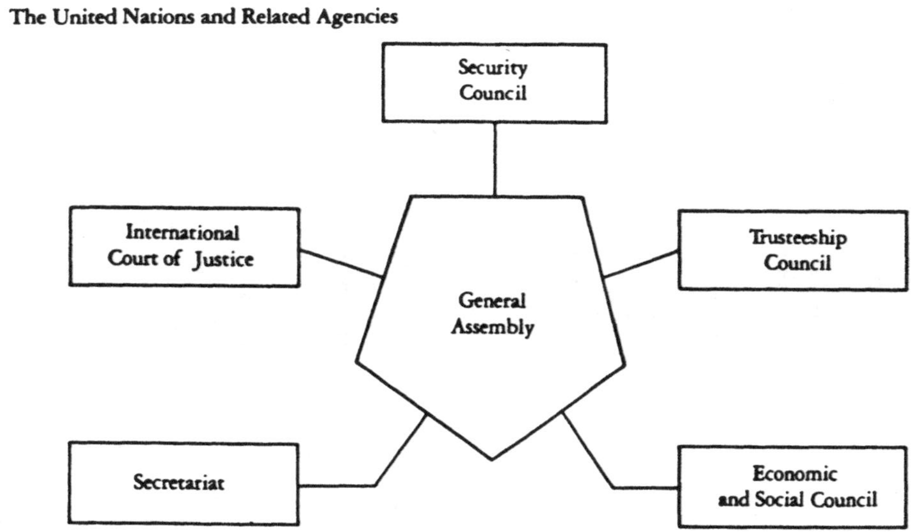

Figure 9.1 The United Nations and Related Agencies
Source: Cheston, T. Stephen, Charles M. Chafer, and Sallie Birket (eds.), *Social Sciences and Space Exploration*, Washington, DC: US Government Printing Office, 1984.

*Related Agencies**

Security Council
Military Staff Committee
Disarmament Commission
United Nations Operations in the Middle East
International Atomic Energy Agency

General Assembly
United Nations Scientific Advisory Committee
Scientific Committee on Effects of Atomic Radiation
Committee on the Peaceful Uses of Outer Space
Committee on Information from Non-Self-Governing Territories
International Law Commission
Advisory Committee on Administrative and Budgetary Questions
Committee on Contributions

Other Subsidiary Bodies of General Assembly:
Disarmament Commission
United Nations Administrative Tribunal
International Atomic Energy Agency
United Nations Emergency Force
United Nations Relief and Work Agency for Palestine Refugees
United Nations Special Fund
United Nations Childrens Fund (UNICEF)
Office of United Nations High Commissioner for Refugees

Secretariat
Administrative Committee on Coordination
Technical Assistance Board

Economic and Social Council
International Atomic Energy Agency
United Nations Special Fund
United Nations Childrens Fund (UNICEF)
Office of United Nations High Commissioner for Refugees
Regional Economic Commissions
Functional Commissions
Administrative Committee on Coordination
Technical Assistance Board

Specialized Agencies:
International Labor Organization
Food and Agriculture Organization of the United Nations
United Nations Educational, Scientific, and Cultural Organization
World Health Organization
International Development Association
International Bank for Reconstruction and Development
International Finance Corporation
International Monetary Fund
International Civil Aviation Organization
Universal Postal Union
International Telecommunication Union
World Meteorological Organization
Inter-Government Maritime Consultative Organization
International Trade Organization

*Some related agencies are listed under more than one UN body; in these cases, the related agencies report to or serve both UN bodies.

Figure 9.1 (cont'd)

Although discussion of space communication began in the ITU as early as 1959, in those early years DBS was only a remote possibility. Still, there was talk at both extremes of the spectrum of discussion regarding direct broadcast satellites. Suggestions ranged from complete prohibition to complete freedom.[7]

France and some lesser developed countries wanted a complete ban of direct broadcast until regulation had been established. The United States proposed that experimentation in certain allocated bands be allowed until study by the ITU could determine appropriate guidelines. At this early time, the Soviet Union and the United States agreed, thus "creating a brief mini-alliance that enabled the ITU to authorize experimental use of satellite broadcasting in a number of narrow frequency bands. In effect with direct broadcast barely on the horizon, the dispute was postponed."[8]

The ITU was further charged with the province of space communication in 1961 when the General Assembly unanimously adopted Resolution 1721 (XVI). The ITU was invited to consider "requests from Member States for technical and other assistance . . . so that they may make effective use of space communications. . . ."[9] The role of the ITU was again dealt with the following year in General Assembly Resolution 1802 (XVII), which was also unanimously adopted.[10]

9.3.1 The 1963 UN Space Law Declaration

The issue of DBS was first raised in the UN-COPUOS by the Brazilian delegate attending a meeting in November 1963.[11] At that time, the Committee was discussing the draft Declaration of Legal Principles Governing the Activities of States in the Exploration and Use of Outer Space (Space Law Declaration). The Brazilian delegate stated:

> The Declaration should also incorporate a ban on the utilization of a communication system based on satellites for purposes of encouraging national, racial or class rivalries.[12]

Thus, an expression of attitudes consistent with those still voiced in the United Nations today, specifically, attitudes that demand a provision requiring the prior consent of receiving States, was first heard in the UN forum over twenty years ago.

While the suggestion from the Brazilian delegation was not incorporated as a provision into the Space Law Declaration, which was passed a month later, the issue was brought to the forefront in the United Nations and demanded attention.[13]

On 13 December 1963 the General Assembly unanimously passed two basic documents that were direct predecessors of the 1967 Treaty on Outer Space: Resolution 1962 (XVIII) on Legal Principles (Space Law Declaration) and Resolution 1963 (XVIII) on International Cooperation.[14] The principles in these two documents were used almost intact for the 1967 Treaty on Outer Space. The 1967 Treaty on Outer Space stands as the basic international instrument in this field.[15]

In Resolution 1963 (XVIII), decisions of the ITU on "the allocation of frequency bands for space communications and procedures for their use as a step in the development of space radio communications," were welcomed.[16] A progress report from the ITU was therein requested. Recall that 1963 was also the year of the Extraordinary Administrative Radio Conference to Allocate Frequency Bands for Space Radiocommunication Purposes — the first conference on space communication.

9.3.2 Initial Consideration of Direct Broadcast Satellites

In 1964 UN-COPUOS reported that it had decided to consider questions relating to the use of satellites for transmitting radio and television programs intended for direct reception by the general public after the report on the subject by the International Radio Consultative Committee (CCIR) of the ITU had been received.[17] In December 1965, in an unanimous decision, the General Assembly adopted Resolution 2130 (II and III) (XX) that welcomed further progress reports from the ITU.[18] In 1965 COPUOS, in its Scientific and Technical Subcommittee, considered the fourth progress report regarding space communication from the ITU.

Direct broadcast satellites had been identified and considered within the United Nations some years before the draft text of the Treaty on Outer Space was adopted by the General Assembly. Twenty-one years earlier, and twelve years before an artificial earth satellite was orbited, Arthur C. Clarke published an article describing communications from outer space to the Earth and commented that "[a] true broadcast service, giving constant field strength at all times over the whole globe would be invaluable, not to say indispensable, in a world society."[19]

9.4 THE 1967 UN TREATY ON OUTER SPACE

On 19 December 1966, after years of arduous work, the General Assembly unanimously adopted the recommendation of COPUOS for a "Treaty on Principles Governing the Activities of States in the Exploration and Use of Outer Space, Including the Moon and Other Celestial Bodies."

It was opened for signature on 27 January 1967 and entered into force on 10 October 1967.[20] This "Outer Space Treaty" or "Space Charter" has been characterized as a Magna Charta for space.

The Outer Space Treaty, although not contemplating direct broadcast specifically by mention, is the foremost source of international space law and is intended to create a framework of principles to guide all activities in the realm of outer space. As direct broadcasting from orbital devices takes place in this realm, the Treaty on Outer Space pertains to DBS activities.

The Outer Space Treaty is considered to be a milestone in the development of international law. The Treaty arose from a process of consensus in COPUOS, where earlier disputes regarding membership and voting had been solved by establishing a large membership bound by the principle of consensus agreement. With original membership of 24 States, COPUOS has been expanded several times, so that by its 35th session in 1980, COPUOS had grown to 53 member states.[21]

By COPUOS procedure, on any perceived agreement, a statement of understanding is read by each Member State into the record. This is to determine to what degree a true consensus exists and to preserve each government's unique interpretation of any existing consensus.

Analysis of the debates, resolutions, and ratifying documents that accompany the Outer Space Treaty confirm its quasi-constitutional intent. The Treaty was designed to create a set of overriding principles that should govern subsequent multilateral and bilateral agreements. The 1967 Outer Space Treaty, which as of 1984 had assembled 90 signatures, accessions, or notifications of successions,[22] embodies many modern and novel concepts and emphasizes the internationalization of outer space. The legal principles that can be derived from the language of the Treaty are:

- International law and the Charter of the United Nations shall apply to space activities.
- Outer space shall be free for exploration, use, and scientific investigation.
- There can be no claims of sovereignty or territory by nations on Earth over locations in space, other than over objects launched from Earth.
- Outer space and celestial bodies are the province of all mankind and should be used only for peaceful purposes.
- Nuclear weapons, weapons of mass destruction, military bases and maneuvers are banned from space.
- The launching state retains jurisdiction over space objects launched from Earth.

- Governments are liable for damage caused on Earth by their space objects.
- Private interests are recognized as having freedom of action in space as long as a government or group of governments on Earth authorize, assume responsibility for, and continually supervise their activities. Signatory nations, therefore, have a duty to oversee the activities of their citizens and commercial space ventures.
- As "Envoys of Mankind," astronauts are due noninterference and all necessary assistance in distress.
- Steps must be taken to ensure that the natural environments of celestial bodies not be contaminated by earth organisms and that the Earth not be contaminated by extraterrestrial organisms.[23]

These, then, are the basic principles that establish the framework within the United Nations for work on a legal regime related to direct broadcast satellites.

9.5 THE ISSUE OF DIRECT BROADCAST SATELLITES

Following an initiative by Canada and Sweden to provide for the establishment of a multidisciplinary forum,[24] the General Assembly, on 20 December 1968, unanimously adopted Resolution 2453 (XXIII), which approved the creation of the Working Group on Direct Broadcast Satellites. (See Figure 9.2 and Table 9.1.) Established by the Legal Subcommittee of COPUOS, the Working Group was "to study and report on the technical feasibility of communication by direct broadcast from satellites . . . as well as the implications of such developments in the social, cultural, legal and other areas."[25]

Since 1968, the Working Group, COPUOS, and its Legal Subcommittee have been the principal fora for intergovernmental deliberations on the implications of DBS. In November 1976, the General Assembly adopted Resolution 31/8 calling for the Subcommittee to consider the task of drafting principles of DBS a matter of high priority. As a result, the Subcommittee established its Working Group II on Television Broadcasting.[26]

As noted earlier, DBS is legally defined as "a radio communication service in which signals transmitted or retransmitted by space stations are intended for direct reception by the general public."[27] A footnote to that definition provides that, "[in] the broadcasting service, the term direct reception shall encompass both individual reception and community reception."[28]

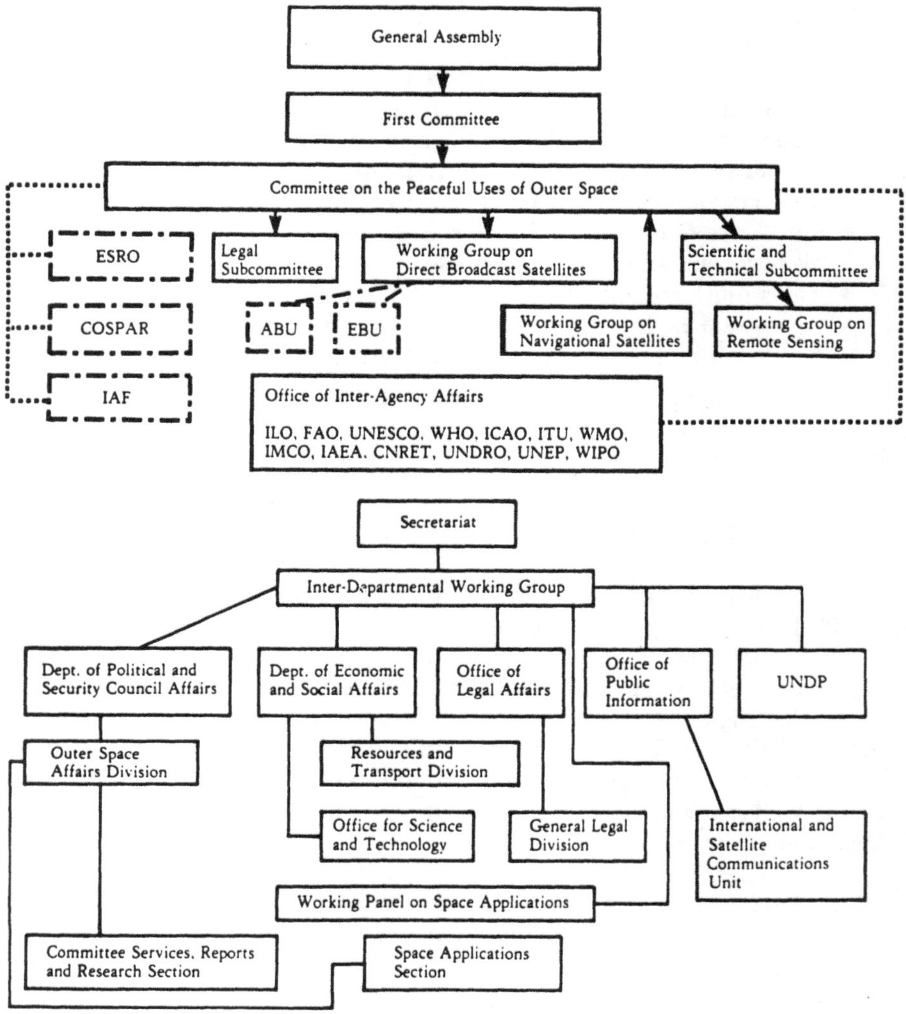

Note: Dotted lines represent observers.

Figure 9.2 International Bodies Concerned with Outer Space UN Bodies
Source: *Proceedings, International Conference on Doing Business in Space*, Smithsonian Institution, American Law Institute, 1981. From presentation by Bert Cowlan.

Table 9.1 Explanation of Acronyms Used in Figure 9.2

Acronym	Explanation
Specialized Agencies	
ILO	International Labor Organization
FAO	Food and Agriculture Organization of the United Nations
UNESCO	United Nations Educational, Scientific, and Cultural Organization
WHO	World Health Organization
ICAO	International Civil Aviation Organization
ITU	International Telecommunication Union
WMO	World Meteorological Organization
IMCO	Inter-Governmental Maritime Consultative Organization
IAEA	International Atomic Energy Agency
Other Inter-Governmental Agencies	
ESRO	European Space Research Organization
INTELSAT	International Telecommunications Satellite Organization
INTERSPUTNIK	International System and Organization of Space Communications
Non-Governmental Organizations	
ICSU	International Council of Scientific Unions
COSPAR	Committee on Space Research of the ICSU
IAF	International Astronautical Federation
ABU	Asian Broadcasting Union
EBU	European Broadcasting Union
Others	
Office of Inter-Agency Affairs	
WIPO	World Intellectual Property Organization
CNRET	U.N. Centre for Natural Resources, Energy and Transport
UNDRO	U.N. Disaster Relief Organization
UNEP	U.N. Environment Programme
Miscellaneous	
UNITAR	U.N. Institute for Training and Research
UNDP	United Nations Development Programme

The television signal coverage of DBS far surpasses the coverage offered by terrestrial television transmitters. Three satellites can virtually blanket the entire Earth with the highest quality of signal. The fact that satellites offer such high quality and availability, and that it is the powerful medium of television that can be transmitted, has caused the international community to recognize the immense potential of DBS, both advantageous and disadvantageous.

While the report of the DBS Working Group in 1969 mainly dealt with the technological feasibility of DBS and certain economic considerations, the report did recognize that the development of DBS "holds the promise of unprecedented progress in communications and understanding between peoples and cultures,"[29] for example, improved health and education and a greater flow of news and information. The report also warned of adverse effects if the new technology was abused.

These potential adverse effects "can be seen as resulting from the receipt of (basically television) programs in the territories of states whose governments do not wish to have them received."[30]

Potential programs can be categorized by content:

- programs aimed at achieving certain potential objectives, such as propaganda, incitement, and interference;
- programs that are inadmissible because they use material or techniques outlawed in the receiving States, such as violence, obscenity, and subliminal advertising;
- programs that are unwanted mostly because they are foreign and allegedly have the potential to undermine national integrity, such as news or public information programs. A good deal of concern arises because so much programming is produced in a single country, that is, the United States of America.[31]

9.5.1 Approaches to Direct Broadcast

The topic of DBS has been on the agenda of the Legal Subcommittee of COPUOS since the creation of the Working Group in 1968. Before that time, discussion on the subject mainly revolved around the technical feasibility of the new technology, however, discussion soon shifted to talk of its legal control.

At the first session of the DBS Working Group all but the United States seemed to want some form of regulation of direct broadcasting. The issue of regulating DBS, both to form and to substance, has created, over the years, a legal impasse on the issue of prior consent of countries to direct broadcast from space. Various approaches have emerged at the international level designed to resolve the conflict. Suggestions for the development of technical procedures, international cooperative arrangements, declarations of guiding principles, and international conventions have been proposed.[32]

In 1973 the Soviet delegation submitted a proposal for a Convention on Principles Governing the Use by States of Artificial Earth Satellites for Direct Television Broadcasting. This approach stressed the need for firm and precise international treaty obligations. The key provision of the Soviet Draft Convention permitted broadcasts "by means of artificial earth satellites to foreign States only with the express consent of the latter."[33]

This prior consent principle has been repeated many times and exists today as the greatest impediment to reaching consensus on principles of a DBS convention. An international treaty often follows a convention within a few years, but in the case of DBS the disagreement on the mandate for prior consent has stalled any consensus. An aspect of equally great import was that nowhere in the Soviet Draft Convention was the principle of the free flow of information mentioned.[34]

Besides the Draft Convention on Principles, the Soviets also presented a resolution calling for COPUOS and its Legal Subcommittee to formulate a binding treaty on DBS. The resolution, like the Draft Convention, initiated heavy debate. Western nations, led by the United States, strongly objected to the rigidity of a binding treaty. These nations preferred a more laissez-faire approach that would limit regulation to the technical committees of the ITU. As a result, the Soviet resolution was amended in COPUOS to call only for an elaboration of principles to govern DBS, with a view to concluding a binding international agreement at some point in the future. The amended resolution (G.A. Res. 2916), a commitment to elaborate DBS principles, passed the General Assembly by a vote of 102 to 1, the United States dissenting.[35]

The Soviets then amended their Draft Convention on Principles in an attempt to take advantage of the new United Nations commitment to formulate such a document. Although somewhat more moderate than the prior draft, the new version of the Soviet Draft Convention still contained the mandate for prior consent.[36] Thus, despite the new United Nations commitment, the new Soviet Draft Convention was as objectionable to most of the West as was their original version of the Draft Convention. Since then the Soviet position has changed very little. The Soviets continue to stress the importance of national sovereignty, and their influence remains important.

In 1972 Canada and Sweden also proposed an approach to resolving the controversial issues of DBS. They suggested a compromise stressing international cooperation in establishing and operating DBS. The approach flows from the belief that it would be premature and possibly counterproductive to try to formulate detailed legal rules and principles before experience had been gained in the establishment and operation of DBS.[37]

In 1972 DBS was not viewed as possible until, at least, 1985. The suggestion was to gain operating experience first, especially through the cooperation of groups of countries in the same regions. It was thought that this cooperation could establish practical patterns of operation and could thus provide an indication of the legal norms that were or were not necessary.[38]

The Canada-Sweden proposal specifically stated that DBS activities to any "foreign State shall be undertaken only with the consent of that State." Moreover, "[t]he consenting State shall have the right to participate in activities. . . ."[39] Yet, Canada and Sweden opposed any expressed rules governing content. They felt that the consent and participation principles rendered content rules unnecessary.[40] This was intended as a compromise aimed toward bringing a convention on principles to consensus.

Also in 1972, the United Nations Educational, Scientific and Cultural Organization (UNESCO), also a Specialized Agency of the UN, proposed yet another approach that sought the development of guiding principles. This approach was reflected in the UNESCO "Declaration of Guiding Principles on the Uses of Satellite Broadcasting for the Free Flow of Information, the Spread of Education, and Greater Cultural Exchange." The UNESCO declaration basically represented a rejection by many developing countries of the "excessively laissez-faire nature" of the Canada-Sweden compromise and was more a regulatory approach that prescribed general and specific principles to govern direct broadcasting.[41]

A final approach, a technical approach carried on largely under the auspices of the ITU, involved elaborating rules on technical characteristics and consultation procedures. As noted in Chapter 6, as a result of the World Administrative Radio Conference for the Planning of the Broadcasting-Satellite Service in the 12 GHz band called by the ITU in 1977, a limit on the transmissions of broadcast satellites was established.

In light of the political, economic, social, and cultural complications of the new DBS technology, a plan was, for the first time, drawn up for space radiocommunication. The plan contained important technical characteristics including the position of the satellite in the geostationary satellite orbit and the region to be covered by television signals from any particular satellite. The plan also elaborated rules regarding consultation procedures that would confine satellite broadcast transmissions to national territories or, in an extremely limited number of cases, to other countries with their concurrence. Out of the 113 countries that attended the conference, accredited representatives from 106 countries signed the Final Acts of WARC-77 on 13 February 1977 and they entered into force on 1 January 1979.[42]

9.5.2 The Legal Impasse Surrounding Direct Broadcast

The approaches discussed in Section 9.5.1 were proposed with an aim to resolving the conflict surrounding DBS. Still, not all of the draft principles governing DBS met with agreement. The negotiators were able to agree, some not so willingly, to principles relating to Purpose and Objectives, Applicability of International Law, Rights and Benefits, International Cooperation, Peaceful Settlement of Disputes, State Responsibility, Duty and Right to Consult, Copyright and Neighboring Rights, and Notification to the United Nations.[43]

Nevertheless, three major issues remained as impediments to a more comprehensive agreement: prior consent, program content, and radiation (signal) overspill. In the 1977 meeting of the DBS Working Group of the Legal Subcommittee of COPUOS, an attempt was made to reconcile the two different basic approaches to the issue of DBS, namely prior consent required by the national sovereign versus free flow of information. The DBS Working Group sought to replace the principle of "Consent and Participation" with a principle on "Consultation and Agreement between States." The Working Group then proceeded to formulate a tentative text of a principle under this heading. This principle has remained since as "Consultations and Agreements between States."[44]

The issue of overspill was incorporated into this principle, thus relegating its regulation to the ITU as was provided for in Paragraph 428A of the 1971 revised Radio Regulations. Program content has been implicitly subsumed under "Consultations and Agreements between States" as for agreement on program content, prior consent must logically take place first. Even though consent would not be required for each program, it would be required for any broadcast service. The position of the United States is that agreement on program content is inseparable from the issue of prior consent for reasons just mentioned. Therefore, the issue of prior consent — implicitly subsumed now under the principle on "Consultations and Agreements between States" — is the one remaining principle that has not met with agreement by the United States and a few other countries and is the one remaining principle that creates the legal impasse over the regulation of DBS.

The United Kingdom, in a working paper submitted to the Legal Subcommittee of COPUOS in March 1979, stated that the question of prior consent had to be radically rethought and suggested that, in light of the allocations of orbits and spectrums at the 1977 WARC-BC, no further regulation was necessary.[45]

Neither this proposal nor the diluted formulation of the Canada-Sweden proposal met with consensus agreement. The former was rejected by the components of national sovereignty and prior consent; the latter was too strong for those favoring the free flow of information.

The issue of prior consent remains still at impasse, essentially unchanged and seemingly unchangeable. Agreement is not foreseen in the near future, the reason: there are two divergent philosophies underlying the positions of different states. Some delegations express the view that DBS activities should be based on respect for the sovereign rights of states and noninterference in internal affairs. Other state delegations feel that DBS activities should be based on:

> . . . the recognition of the importance of the right of everyone to freedom of expression including the right to seek, receive and impart information and ideas regardless of frontiers, as enshrined in the instruments of the United Nations relating to universal human rights, and that this principle should be contained in the text of principles to be adopted.[46]

The advocates of the principle of prior consent argue that the state has the right freely to select and to develop the unique matrix characterizing its social, political, economic, and cultural system. Those in opposition feel that to mandate prior consent would, in essence, legitimize international censorship and stifle technological progress.

> In summary form, the USA position can be described as a minimalist and common-law-inspired attitude. If any principles should be adopted at all, they could possibly be given the form of a nonbinding declaration. They should be very general, be mainly mindful of principles of free flow and freedom of information, and could include a principle of prior consultation. A principle of prior consent by a "receiving country" has consistently been resisted. The USSR position can be described as a maximalist, civil law conditioned position. Legal principles should be laid down in a binding treaty, which was originally proposed to include specific sanctions. Apart from a principle of prior consent, there should also be detailed content rules and provisions for measures concerning "illegal" broadcast.[47]

The position of the United States in support of the free flow of information is definitely the minority position. But because COPUOS and its subcommittees proceed by the principle of consensus, the principle of free flow is protected from being overruled in either COPUOS or in its DBS Working Group.

In Resolution 36/35 of 3 December 1981, the General Assembly decided that the work of completing the elaboration of a draft set of principles governing DBS should be done by COPUOS itself. The Committee was to deal with the issue during its session in 1982 and then the General Assembly would consider the adoption of such a set of principles at its 37th session in the same year. Since the United Nations had committed itself to enunciate principles nine years earlier (G.A. Res. 2916), the developing countries served notice in 1981 that if no consensus could be reached in the 1982 meeting of COPUOS, they would draft a convention on principles acceptable to a majority of the world's nations and press it to a vote in the General Assembly.[48]

Finally, after several attempts to reach a compromise in COPUOS had failed, the United Nations General Assembly adopted Resolution 37/92 on 10 December 1982. Included in its Annex are Principles Governing the Use by States of Artificial Earth Satellites for International Direct Television Broadcasting.

The draft of this resolution was sponsored by a group of developing countries and the resolution was indorsed by a large majority of Member States. "However, for the first time in the history of space law in the United Nations, this document was not adopted by consensus."[49]

Resolution 37/92 of 10 December 1982 was adopted by a recorded vote of 107 votes in favor, 13 votes against, and 13 abstentions, with 23 member states absent. Later, one of the absent member states announced it had intended to vote in favor.[50] Although the adoption of draft principles often precedes a treaty by a few years, under the circumstances, where the vote was pressed and no consensus was reached, it is doubtful that a binding treaty on DBS will be ratified in the near future.

9.6 CONCLUSION

Even an organization like the ITU is only one part of a larger international system in our complex, global village. As a specialized agency of the United Nations, the ITU has recognized the limited role that it plays in the overall enunciation of international law. The ITU also has recognized that direct broadcasting, specifically, and space communication, generally, fall within the larger sphere of general principles of international space law.

In a reciprocal fashion, the United Nations has recognized the special competence and jurisdiction of the ITU and has repeatedly called upon the ITU for reports and for regulation. In the matter of DBS, the technical approach of the ITU has resulted in coordination and planning procedures, involving something close to a *de facto* form of prior consent, that has been agreed to even by the United States.

Yet, in the more political UN forum, with larger principles of law and human rights at stake, the legal impasse that surrounds the matter of DBS continues to result from the last remaining point of contention: the principle of Consultations and Agreements between States, more specifically, the issue of prior consent. This impasse will always remain, as long as there are countries willing to ignore and others willing to insist upon the principle of Article 19 of the Universal Declaration of Human Rights, that:

> Everyone has the right to freedom of opinion and expression; this right includes freedom to hold opinions without interference and to seek, receive and impart information and ideas through any media and regardless of frontiers.

REFERENCES

1. *United Nations General Assembly Resolution 1348 (XIII),* December 13, 1958.
2. *Ibid.*
3. *United Nations General Assembly Resolution 1472 (XIV),* December 12, 1959.
4. *Ad Hoc* Committee on the Peaceful Uses of Outer Space. *Report. Official Records of the General Assembly, Fourteenth Session, 1959, Annexes,* (United Nations Document A/4141).
5. Ploman, Edward W., "Satellite Broadcasting, National Sovereignty and Free Flow of Information," *National Sovereignty and International Communication,* (K. Nordenstreng & H. Schiller eds.), Norwood, NJ: Ablex, 1979, p. 157.
6. Parry, A., *Russia's Rockets and Missiles,* New York: Doubleday, 1960, p. 208.
7. Freeman, Jennifer, "Direct Broadcast Developments and Directions: The National Sovereignty and Cultural Integrity Positions," *Proceedings of the Seventy-Fourth Annual Meeting,* Washington, DC: American Society of International Law, 1981, pp. 302-321, at 303.
8. *Ibid.*
9. *United Nations General Assembly Resolution 1721, (XVI),* December 20, 1961.
10. *United Nations General Assembly Resolution 1802 (XVII),* December 14, 1962.
11. Dalfen, Charles M., "Principles Governing Direct Satellite Broadcasting," *Manual on Space Law,* Vol. 1, Dobbs Ferry, NY: Oceana Publications, 1979, pp. 283-302, at 283.
12. *Ibid.*
13. *Ibid.*
14. *United Nations General Assembly Resolution 1962 (XVIII),* December 13, 1963; *United Nations General Assembly Resolution 1963 (XVIII),* December 13, 1963.
15. Treaty on Principles Governing the Activities of States in the Exploration and Use of Outer Space, Including the Moon and Other Celestial Bodies; 18 UST 2410, TIAS No. 6347, 610 UNTS 205 No. 8843 (signed at Washington, London, and Moscow, January 27, 1967).
16. *United Nations General Assembly Resolution 1963 (XVIII),* December 13, 1963.

17. *Ad Hoc* Committee on the Peaceful Uses of Outer Space. *Report. Official Records of the General Assembly, Nineteenth Session, 1964-65. Annexes,* Vol. 2 (United Nations Document A/5785).
18. *United Nations General Assembly Resolution 2130 (XX),* December 21, 1965.
19. Clarke, Arthur C., "Extraterrestrial Relays, Can Rocket Stations Give World-wide Coverage?" *Wireless World,* 1945.
20. See *supra* note 15.
21. Kopal, Vladimir, "Evolution of the Main Principles of Space Law in the Institutional Framework of the United Nations," *Journal of Space Law,* Vol. 12, 1984, pp. 12-25, at 13.
22. *Ibid.*, p. 16.
23. See *supra* note 15.
24. Ploman, *supra* note 5, at 159.
25. *United Nations General Assembly Resolution 2453 (XXIII),* December 20, 1968.
26. *United Nations General Assembly Resolution 31/8,* November 12, 1976.
27. *Final Acts of the World Administrative Radio Conference for Space Telecommunications, Geneva, 1971,* Para. 84AP Spa 2.
28. *Ibid.*, Para. 84AP.1 Spa 2.
29. *United Nations Document A/AC.105/50.*
30. Dalfen, *supra* note 11, at 284.
31. *Ibid.*, p. 285.
32. For instance, see *Final Acts of the World Administrative Radio Conference for Space Telecommunications, Geneva, 1971,* Para. 428A, and *Final Acts of the World Administrative Radio Conference for the Planning of the Broadcasting-Satellite Service in Frequency Bands 11.7-12.2 GHz (in Regions 2 and 3) and 11.7-12.5 GHz (in Region 1),* Geneva, 1977, (herein cited as Final Acts WARC-BS).
33. Kopal, *supra* note 21, at 19.
34. Freeman, *supra* note 7, at 304.
35. *Ibid.*
36. *Ibid.*
37. Dalfen, *supra* note 11, at 285.
38. *Ibid.*, p. 286.
39. *United Nations Document A/AC.105/WG.3/L.4,* 1973.
40. Dalfen, *supra* note 11, at 285.
41. *Ibid.*

42. Butler, Richard E., "World Administrative Radio Conference for Planning Broadcasting Satellite Service," *Journal of Space Law,* Vol. 5, 1977, pp. 93-99, at 94; also Final Acts WARC-BS, *supra* note 32.
43. For a clear and concise historical discussion on each of these principles see Dalfen, *supra* note 11, at 285.
44. Kwen, Chen, "Pending Issues Before the Legal Subcommittee of the United Nations Committee on Peaceful Uses of Outer Space," *Journal of Space Law,* Vol. 5, 1977, pp. 29-35, at 31.
45. Dalfen, *supra* note 11, at 293.
46. "Report of the Chairman of the Working Group on Direct Broadcast Satellites," *Journal of Space Law,* Vol. 8, 1980, pp. 65-70, at 66.
47. Ploman, *supra* note 5, at 160.
48. Jasentuliyana, Nandasiri, "The Work of the UN Committee on the Peaceful Uses of Outer Space in 1982," *Journal of Space Law,* Vol. 10, 1982, pp. 41-46, at 44.
49. Kopal, *supra* note 21, at 20.
50. *Ibid.*

Conclusion

Space Communication: Toward Cultural and Legal Convergence in the Global Village

Twenty-five years have elapsed since the first communication satellite provided a communication link from the geostationary satellite orbit. Launched in 1963, the US National Aeronautics and Space Administration's Syncom II was used for many experiments, including the transmission in 1964 of live television from the Tokyo Olympic Games.[1] Since that time, geostationary communication satellites have transformed world communication, serving as important links in domestic and international telecommunication networks. Geostationary communication satellites are among the most effective and economical telecommunications systems available.

Since the mid-1960s, the demand for geostationary satellite communication has increased dramatically as these satellites have become indispensable instruments in domestic and international communication of all kinds. Indeed, in some developing nations, satellites bring communication and culture to towns where telephone trunks do not even exist.

With this increase in demand for geostationary satellite communication has come an increase in demand for the two resources necessary to make this communication possible — radio frequency spectrum and space in the geostationary satellite orbit, commonly referred to as the spectrum-orbit resource. Although characterized as a limited natural resource, the capacity of the spectrum-orbit resource has been significantly extended as a result of technological progress, which has allowed for more effective use of the radio spectrum and the GSO. With the use of resource enhancement techniques like frequency re-use, signal polarization, beam shaping, reduced satellite spacing, improved earth station technology, and a multitude of other techniques and technologies, the capacity of the GSO to accommodate communication satellites has been substantially increased.[2]

Many developing countries, however, are concerned that the capacity of the spectrum-orbit resource may soon be reached. These countries are particularly concerned that the most desirable, or even acceptable, orbital slots will not be available by the time they are able to use them. In the mid-1970s, a country could obtain orbital slots for communication satellites from the ITU almost routinely. By the end of the decade, obtaining acceptable slots in the GSO used to service North America and the USSR began to require more attention to technical characteristics and more cooperation between nations was necessary to allow for an effective choice.[3]

Because of the integral relation between the use of space and the use of radiocommunication, the first actual enactments of international space law were laid down in the International Telecommunication Convention and its annexed Radio Regulations of 1959. Even though a General Assembly Resolution had considered legal principles for space law a year earlier, it was not until the ITU revised its Convention and Regulations that the first laws for space were actually established. Indeed, the enormous potential of space communication to restructure social and economic patterns so as to create a global village impelled the countries of the world to consider the accommodation and regulation of space communication in the ITU eight years prior to the first UN space law treaty — the Outer Space Treaty of 1967.

Canadian communication theorist Marshall McLuhan forecast the global village in the late 1950s. Arthur C. Clarke defined the technical aspects and potential of geostationary communication satellites in his famous *Wireless World* article as early as 1945. The Bell Telephone Laboratories and others advanced Clarke's theory to the point of technical demonstrations toward the late 1950s and early 1960s, with projects like SCORE, ECHO, ADVENT, TELSTAR, RELAY, SYNCOM, and culminating in 1965 with INTELSAT's Early Bird — the world's first commercial communication satellite.[4] As this technical progress moved the world more rapidly toward a global village, the ITU began to develop the necessary technical and legal approaches to allow for progress to be made in an orderly and effective manner.

Satellite communication has indeed served as the nexus, the information connection, the catalyst for critical mass in a growing technological, cultural, and legal convergence in the global village. This convergence has been characterized as the "information revolution," the "information age," the "second industrial revolution," the "third wave," the "space age," and various other names. Satellite technology has helped to bring about this cultural implosion, which is having profound and unforeseen consequences. The very ability of humanity to cope with the diversity, the passionately held beliefs, and the remarkable technology characteristic of this time is being tested, both in international organizations like the ITU and the United Nations and in the living rooms and social centers of our watching and listening world.

It is for these reasons that the ongoing ITU exercise in international cooperation is so important. The unrest in the ITU today is merely symptomatic of the momentous change occurring in the world. Just as the relatively developed half of the world was emerging from a first century of industrial revolution into the space and information ages, the rest of the preindustrial world was thrust headlong into both industrial and information revolutions simultaneously, with little or no preparation or awareness of the consequences.

No region of the world, no matter how previously developed, is immune to the effects of this epochal change. In all the eons of time, so far as we know, there has been no other era when such great numbers of sentient creatures on this planet have been instantaneously and simultaneously aware of one another's basic day-to-day circumstances. Marshall McLuhan's famous aphorism "the medium is the message" forecast the inherent potential for revolution and evolution in global, instantaneous, electronic communication — a new village of human intercourse would be formed, one that would require global arrangements and global institutions.

The message of electronic and space communication is realized in the international negotiations that "unofficially" take place on televised interview shows and in statements by public officials. The message of the media is evident in the modern hostage crises, which are almost totally dependent for their effect upon the electronic media. We experience the message again when private satellites broadcast the explosion of a nuclear power plant in one of the most closed societies on Earth — pictures broadcast by the constitutionally protected news media in the United States.

The occurrence of the African famine in the mid-1980s that resulted in a worldwide effort by entertainment and media interests to deliver help are indicative of the broad, evolutionary reach of global, instantaneous space communication. We know, as if we were neighbors, the expressions half a world away of starving children, raging terrorists, religious potentates, and even world leaders in the midst of nuclear negotiations as they exchange words of conflict or cooperation.

The impact of satellite technology is evident also from the standpoint of cultural and technological development, as it has become an important tool for the realization of economic and social goals through various applications of space communication, including satellite communication, earth exploration, meteorology, maritime mobile communication, radiodetermination, space research, and other applications. We now hear of "integrated services digital networks" (ISDN) and of the possibility of the ITU becoming an International Teleinformatics Union. We hear that space-based strategic defense, or "Star Wars" is fundamentally dependent upon the capacity, defensability and vulnerability of its command and control systems, its teleinformatic nervous systems. It is evident that the world has been permanently changed.

It should not be surprising that the prime international agency for coordinating and regulating international telecommunication could be profoundly changed through the fulfillment of its responsibility for overseeing this technocultural behemoth called space communication. Space communication has profoundly challenged the prevailing model or approach to the regulation of telecommunication, specifically radiocommunication — a model that has existed since the first radio conferences as far back as 1906.

The impact of space communication, however, transcends the mere movement in the ITU toward *a priori* kinds of planning for some of the space communication services. It transcends the creation of development assistance capacity in the ITU, an event directly related to the interaction of space communication with the simultaneous change of cultural epoch occurring in more than one-half of the world. The influence of space communication on the international telecommunication environment has resulted in the development of international legal principles based upon the ethical principle of equity.

These legal models were designed to help define the meaning of equity, or at least to suggest some of the prerequisites necessary for its existence in an important, even fundamental, human activity like international communication. The development of models for effective and equitable management of competition for access to space resources could prove to be of enormous importance in a future sure to see the development of increasing competition and conflict in the arena of space activities and in the management of global resources of all kinds.

The legal evolution and restructuring that has been examined in this book is but one demonstration of what many theorists have postulated — that instantaneous, worldwide communication inherently contains the seed for an evolutionary impact on global culture, society, politics, and law. The evolution of technical assistance for developing countries in the ITU in response to the growing demand for equity is confirmation of this impact. The growth of technical assistance in the ITU is evidenced by the adoption of several provisions over time, from the amendment of Article 4 in the 1959 ITU Convention to foster ". . . the creation, development and improvement of telecommunication equipment and networks in developing countries by every means at its disposal . . .," to the inclusion of eight resolutions and one opinion in the 1965 Convention, ten resolutions and one opinion in the 1973 Convention, and fifteen resolutions in the 1982

Convention — all aimed at providing technical assistance to the developing countries for improvement of their telecommunication capabilities.

Technical assistance is a practical or procedural approach to accommodating the demand for access to the resources for space communication. The evolution of the principle of equity in the ITU is a reformulation of substantive international law. This reformulation began with the adoption of Recommendation 10A at the 1963 Space EARC. Recommendation 10A was an expression of foresight that was followed and elaborated upon by Resolution 24 of the 1965 Plenipotentiary, Resolution Spa 2-1 of WARC-ST-71, Resolution 2 of the 1979 GWARC, and Article 33 of the 1973 and the 1982 International Telecommunication Conventions. (See Table C.1.)

Still remaining to be completed is that which is the most difficult to achieve — the application of principle to practice in the late 1980s and 1990s, and beyond into the twenty-first century. Though great and unforeseen changes may or may not be ahead for the constitution or structure of the ITU, this profitable and thus far successful exercise in international human communication and cooperation is likely to continue, because cooperation is indispensable if radiocommunication is to be effectively, efficiently, and equitably used for the benefit and advancement of all humankind.

As with previous exercises in international cooperation in telegraphy, radiotelegraphy, and telecommunication over more than a 100-year period, the recent transformations in space communication law and regulation afford some reason for optimism and hope concerning global society's navigation of the unforeseeable tides of the future. As we consider the prospects for the future, not only in broadcasting, telephonic, and digital communication, but also in newly emerging areas of mobile communication via satellite, satellite radiodetermination, and the expanding uses of remote sensing satellites, it is perhaps helpful to recall the concluding remarks of Chairman D. Goedhuis of the International Law Association's New Delhi Space Law Committee in the committee summary on "Telecommunications by Satellite:"

> There are . . . *in space developments two factors,* namely the *extraordinary interdependence* in space activities and the *growing convergence* of interests, which *have radically changed the dimensions and objectives of the purely national interest.* . . . One need not be an incorrigible illusionist to expect that the dawning awareness of this interdependence and convergence of interests will growingly make itself felt in regional developments in the field of communication satellites which should be considered as a necessary stage in the *evolution towards globally applicable rules* . . . There are . . . in the case of space, both psychological and social factors at work which militate against an over-extension of national claims.[5]

Table C. 1 Chronological Evolution of Legal Principles and Provisions for Space Communication*

1959 WORLD ADMINISTRATIVE RADIO CONFERENCE	
Article 1	Revised to include two new services: space service and earth-space service.
Article 5	1% of radio spectrum allocated for space services.
1959 GENEVA PLENIPOTENTIARY CONFERENCE	
Article 4	Amended to include as a duty of the Union the fostering of technical cooperation, development, and improvement of telecommunication equipment and networks in developing countries.
Article 13	International Consultative Committees to pay due attention to formulation of recommendations concerning improvement of telecommunications in developing countries.
1963 EXTRAORDINARY ADMINISTRATIVE RADIO CONFERENCE TO ALLOCATE FREQUENCY BANDS FOR SPACE RADIOCOMMUNICATION PURPOSES	
Article 1	Revised to include two categories of satellite communication: Communication-Satellite Service and Broadcasting-Satellite Service (BSS).
Article 5	15% of radio spectrum allocated for space communication on a shared or exclusive basis with terrestrial services.
Article 9A	Procedures established for obtaining international recognition of spectrum uses for space communication.
Resolution 1A	Coordination procedure limited to international satellite systems established.
Recommendation 10A	*Principle of equitable and rational use of radio spectrum established.*
1965 MONTREUX PLENIPOTENTIARY CONFERENCE	
Resolution 24	*Reiterated the applicability of the principle of nondiscriminatory, equitable access to radio spectrum.*

Table C.1 (cont'd)

Technical Cooperation	Eight resolutions including:
Resolution 27	Participation by the Union in UN Development Program;
Resolution 28	Called for methods of improving technical cooperation;
Resolution 29	Called for recruiting of specialists to assist developing countries and for improvement of Union facilities for providing information and advice to these countries.
1971 WORLD ADMINISTRATIVE RADIO CONFERENCE FOR SPACE TELECOMMUNICATIONS	
Article 1	Fixed-Satellite Service and Mobile-Satellite Service replace Communication-Satellite Service; BSS expanded; Earth-Exploration and Maritime-Mobile Satellite Services (MMSS) defined, and frequencies allocated for their use—MMSS facilitates establishment of INMARSAT.
Article 5	About 15,000 MHz of new spectrum allocated to space services.
Article 9A	Revised to include mandatory procedure of advance publication of information on planned satellite systems; to take place prior to coordination.
Article 7, Para. No. 428A	Obligation to minimize radiation overspill of BSS signals.
Appendix 1B	"Advance Publication Information to be Furnished for a Satellite Network" sets forth basic operational data to be provided IFRB on any planned satellite system.
Resolution Spa 2-1	Recognized spectrum-orbit as natural resources, thus linking them; emphasized efficient and economical use; stated countries have equal rights to use spectrum-orbit resources
Recommendation Spa 2-1	*Affirmed proposition that countries should be able to establish their necessary telecommuni-*

Table C.1 (cont'd)

	cation links, and focused on efficient and economical use of the radio spectrum.
Resolution Spa 2-2	Called for *a priori* planning of the BSS.
Resolution Spa 2-3	Established detailed coordination procedure for BSS with conditions for notification, examination, and registration in the IFRB Master Register of BSS assignments.
1973 MÁLAGA-TORREMOLINOS PLENIPOTENTIARY CONFERENCE	
Article 33	*Principle of nondiscriminatory equitable access to spectrum-orbit resource established as formal international law with emphasis on efficient and economical use.*
Article 10	Amended to charge IFRB with greater regulatory responsibilities for space communication, including recording of satellite positions in GSO and advising Members on its efficient and economical use.
Resolution 27	Called for WARC-BS to plan BSS.
Resolution 28	Called for WARC-79 to consider general revision of Radio Regulations.
Technical Cooperation	Ten Resolutions including:
Resolution 19	Special measures for least developed countries to meet their communication needs;
Resolution 21	Special fund for technical cooperation for developing countries;
Opinion 2	Favorable treatment in areas of communication for developing countries.
1977 WORLD ADMINISTRATIVE RADIO CONFERENCE FOR THE PLANNING OF THE BROADCASTING-SATELLITE SERVICE	
Article 11	Established *a priori* allotment Plan for the BSS in Regions 1 and 3.
Resolution Sat-8	Called for RARC for equitable planning of BSS in Region 2.

Table C.1 (cont'd)

Resolution Sat-9	Identified goal of future RARC as drafting of plan for BSS.
Recommendation Sat-8	Elaborated on equitable planning of Region 2 BSS by identifying minimum number of frequency channels for each administration.
1979 WORLD ADMINISTRATIVE RADIO CONFERENCE	
Resolution 2	*Replaced 1971 Resolution Spa 2-1 reiterating no permanent priority to registration of frequency assignments.*
Resolution 3	*Called for guarantee in practice for all countries of equitable access to GSO and frequency bands for space services.*
Resolution 101	Called for planning of feeder links in the BSS for Regions 1 and 3.
Resolution 701	Replaced 1977 Resolutions Sat-8 and Sat-9 calling for planning of BSS in Region 2.
1982 NAIROBI PLENIPOTENTIARY CONFERENCE	
Article 33	*Amended to reflect principles of efficiency and economy in light of special needs of developing countries.*
Article 10	Amended so as no longer to require IFRB to record date, purpose, and technical characteristics when registering a frequency assignment.
Article 15	Amended financial contributory structure to allow developing countries to contribute less to financing the Union.
Resolution 1	Called for 13 future conferences, including a 1989 Plenipotenitary and three administrative conferences dealing with space communication: RARC-83, WARC-ORB-85, and WARC-ORB-88.
Technical Cooperation	Eighteen Resolutions including:
Resolution 26	Called for stronger ITU regional presence in developing countries.

Table C.1 (cont'd)

1983 REGIONAL ADMINISTRATIVE RADIO CONFERENCE FOR THE PLANNING OF THE BROADCASTING-SATELLITE SERVICE IN REGION 2	
The Plan	Adopted *a priori* plan for BSS in Region 2, characterized by flexibility.
1985 WORLD ADMINISTRATIVE RADIO CONFERENCE ON THE USE OF THE GEOSTATIONARY-SATELLITE ORBIT AND THE PLANNING OF THE SPACE SERVICES UTILIZING IT	
Appendix 30	Incorporated into Radio Regulations provisions and associated Plan for BSS in Region 2
Appendix 30A	Incorporated into Radio Regulations provisions and associated Plan for feeder links for BSS in Region 2
Guarantee of Access	-An orbital position within a predetermined arc; -A minimum bandwidth; -A service area limited to national systems providing domestic services.
Planning Method	-Arc allotment plan for FSS in expansion bands; -Improved coordination procedures in conventional bands not subject to allotment: multilateral planning meetings.
Planning Principles	-Guarantee of access; -Sharing with other services; -Reservation of resources; -Technical aspects of special geographical situations; -Consideration of existing systems; -Provisions for multiadministration systems; -Flexibility; -Different planning solutions in different circumstances; -Efficiency; -Provisions for multiservice and multiband networks;

Table C.1 (cont'd)

	-Cost of development and application as low as possible.

**Note:* Italics refer to legal principles concerning equity.

REFERENCES

1. Martin, James, *Communications Satellite Systems,* Englewood Cliffs, NJ: Prentice-Hall, 1978, p. 153.
2. "Efficient Use of the Geostationary Orbit," *United Nations Document A/CONF,101/BP/7,* January 16, 1981, p. 20.
3. *First Report of the Advisory Committee for the ITU's World Administrative Radio Conference on the Use of the Geostationary-Satellite Orbit and the Planning of the Space Services Utilizing It,* December 1983, pp. 4-20 to 4-25.
4. See generally, Martin, *supra* note 1, for history of early satellites.
5. Goedhuis, D., "Telecommunications by Satellite," *The International Law Association Report of the Fifty-Sixth Conference Held at New Delhi,* London: Cambrian News (Aberystwyth), 1976, p. 482.

Glossary

Allocation Entry in the Table of Frequency Allocations of a given frequency band for the purpose of its use by one or more terrestrial or space radiocommunication services or the radio astronomy service under specified conditions. This term shall also be appplied to the frequency band concerned.

Allotment (of a radio frequency or radio frequency channel): Entry of a designated frequency channel in an agreed plan, adopted by a competent conference, for use by one or more administrations for a terrestrial or space radiocommunication service in one or more identified countries or geographical areas and under specified conditions.

AM Amplitude Modulation.

Assignment (of a radio frequency or radio frequency channel): authorization given by an administration for a radio station to use a radio frequency or radio frequency channel under specified conditions.

BSS Broadcasting-Satellite Service. Often referred to as direct broadcast satellites or DBS.

C Band 6/4 GHz. More specifically: 5925-6425 MHz and 3700-4200 MHz.

CCIR *Comité consultatif international des radiocommunications*/ International Radio Consultative Committee: a permanent organ of the ITU where Member States and recognized private operating agencies formulate recommendations concerning technical and operational radio matters.

CCIR SPM Special Preparatory Meeting: convened by the Secretary-General of the ITU to provide technical support to an administrative conference.

CCITT *Comité consultatif international télégraphique et téléphonique*/ International Telegraph and Telephone Consultative Committee: a permanent organ of the ITU where Member States and recognized private operating agencies formulate recommendations concerning technical, operational, and tariff aspects of telecommunication.

Comsat Communications Satellite Corporation.

COPUOS Committee on the Peaceful Uses of Outer Space, United Nations.

CPM Conference Preparatory Meeting.

CSAGI Special Committee for the International Geophysical Year.

DBS See BSS, above.

dBW Decibel Watts; a measure of power, decibels referenced to one Watt.

Digital Transmission A technique that transmits the signal in the form of a discrete number of codes. The information content of the signal is concerned with discrete states of the signal, such as the presence or absence of a voltage, a contact in the open or closed position, or a hole or no hole in certain positions on a card.

DSI Digital Speech Interpolation.

EARC Extraordinary Administrative Radio Conference.

EIRP Effective Isotropic Radiated Power.

EESS Earth-Exploration Satellite Service.

EHF Extremely High Frequencies: 30-300 GHz.

Eutelsat European Telecommunication Satellite Organization.

FCC Federal Communications Commission.

FDM Frequency Division Multiplex.

FDMA Frequency Division Multiple Access.

Feeder Link A radio link from an earth station at a specified fixed point to a space station, or *vice versa*, conveying information for a space radiocommunication service other than for the fixed-satellite service.

FSS Fixed-Satellite Service.

FM Frequency Modulation.

Geostationary Satellite A geosynchronous satellite whose circular and direct orbit lies in the plane of the Earth's equator and which thus remains fixed relative to the Earth; by extension, a satellite that remains approximately fixed relative to the Earth.

Geosynchronous Satellite An earth satellite whose period of revolution is synchronized with the period of rotation of the Earth about its axis.

GHz Gigahertz (10^9 Hertz).

GSO The geostationary earth orbit. The orbit is 22,300 miles in space in the equatorial plane.

GWARC General World Administrative Radio Conference.

Harmful Interference Interference that endangers the functioning of a radio-navigation service or of other safety services or seriously degrades, obstructs, or repeatedly interrupts a radiocommunication service operating in accordance with ITU Regulations.

HDTV High Definition Television.

Hertzian Waves See Radio Waves, below.

HF High Frequencies: 3-30 MHz.

Hz Hertz.

ICSC Interim Communication Satellite Committee.

IFRB International Frequency Registration Board; a permanent organ of the ITU with five officials elected by the plenipotentiary conference that examines notifications of frequency assignments from Member States of the ITU for conformity with the Radio Regulations.

IFL International Frequency List; a listing of all the frequencies in use in the world, as notified by administrations to the ITU; printed from information recorded in the Master International Frequency Register and published at intervals not exceeding two years; kept up-to-date by quarterly recapitulative supplements. See IFR, below.

IFR International Frequency Register; (Master International Frequency Register) the official "central file" for all recorded frequency assignments.

IGY International Geophysical Year.

Inmarsat International Maritime Satellite Organization.

Insat Indian National Satellite System.

Intelsat International Telecommunications Satellite Organization with 112 member states as of 1987 that own and operate the satellites in the Global Communication Satellite System.

Interference The effect of unwanted energy due to one or a combination of emissions, radiations, or inductions upon reception in a radiocommunication system, manifested by any performance degradation, misinterpretation, or loss of information that could be extracted in the absence of such unwanted energy.

ISDN Integrated Services Digital Network.

ITU International Telecommunication Union; a specialized agency of the United Nations with responsibilities in the field of international telecommunications including spectrum management.

ITU Convention The governing instrument of the ITU that sets forth its structure and activities; only the plenipotentiary conference of the ITU can amend or revise the Convention.

Ka Band 30/20 GHz.

kHz Kilohertz (10^3 Hertz).

Ku Band 14/11-12 GHz: More specifically Region 1: 11.7-12.5 GHz; Region 2: 12.2-12.7 GHz, and Region 3: 11.7-12.2 GHz.

LDC Lesser Developed Country.

LF Low Frequencies: 30-300 kHz.

MIFR Master International Frequency Register. See IFR, above.

MF Medium Frequencies: 300-3000 kHz.

MHz Megahertz (10^6 Hertz).

MMSS Maritime Mobile-Satellite Service.

MSS Mobile-Satellite Service.

NASA National Aeronautics and Space Administration.

NATO North Atlantic Treaty Organization.

NIIO New International Information Order.

Orbit The path, relative to a specified frame of reference, described by the center of mass of a satellite or other object in space subjected primarily to natural forces, mainly the force of gravity.

OST Outer Space Treaty. The 1967 Treaty on Principles Governing the Activities of States in the Exploration and Use of Outer Space, including the Moon and Other Celestial Bodies.

Palapa Indonesian national satellite system.

PFB Provisional Frequency Board; established by the 1947 Atlantic City Administrative Radio Conference to draft a new International Frequency List for the bands between 4 MHz and 27.5 MHz engineered to the needs of countries as identified at scheduled conferences. The objective of the PFB to "engineer" the radio spectrum was never completely realized, and in 1951 the PFB was dissolved.

pfd Power Flux Density; see below.

Plenipotentiary Conference The supreme body of the ITU that has the power to amend or revise the ITU Convention.

Power Flux Density A measure of the power radiated by a transmitter, used as a constraint on certain services to protect other services in a shared band.

Primary Service A class of allocation. Stations in a primary service may not cause harmful interference to stations in the same, or another primary service, and can claim protection from interference from stations in primary, permitted, and secondary services.

Radio Waves Electromagnetic waves of frequencies arbitrarily lower than 3000 GHz, propagated in space without artificial guide.

RARC Regional Administrative Radio Conference.

RARC SAT-83 Regional Administrative Radio Conference for the Planning of the Broadcasting-Satellite Service in the Frequency Band 12.2-12.7 GHz and Associated Feeder Links in the Frequency Band 17.3-17.8 GHz in Region 2, 1983.

RARC SAT-R2 See RARC SAT-83.

Regions of the ITU For the allocation of frequencies, the world has been divided into three regions by the ITU. Exact boundaries of the regions are given in Article 8 of the Radio Regulations; a general description follows: (Region 1: Europe, Africa, the USSR, the Middle East and Mongolia); (Region 2: North, Central, and South America, the Caribbean, and Greenland); (Region 3: Asia, Oceania, Australia, and New Zealand).

RF Radio Frequency. See Radio Waves, above.

RPOA Recognized Private Operating Agency.

Satellite A body that revolves around another body of preponderant mass and that has a motion primarily and permanently determined by the force of that other body.

Satellite Link A radio link between a transmitting earth station and a receiving earth station through one satellite.

Satellite System A space system using one or more artificial earth satellites.

SCPC Single channel per carrier.

Secondary Service A class of allocation. Stations in a secondary service may not cause interference to stations in a primary or secondary service, and may not claim protection against interference from stations in a primary service existing or subsequently installed.

SHF Super High Frequencies: 3-30 GHz.

SIO Scientific or Industrial Organization.

SITE Satellite Instructional Television Experiment.

SJM Special Joint Meeting of ITU consultative committee study groups.

Space WARC World Administrative Radio Conference on the Use of the Geostationary-Satellite Orbit and the Planning of the Space Services Utilizing It (WARC-ORB).

SPS Solar Power Satellite.

Table of Frequency Allocations A table in Article 8 of the ITU Radio Regulations that specifies which radio services are to use which frequency bands both universally and in regions of the world as designated by the ITU.

TDM Time Division Multiplex.

TDMA Time Division Multiple Access.

TDRS Tracking and Data Relay Satellite.

UHF Ultra High Frequencies: 300-3000 MHz.

UNESCO United Nations Educational, Scientific and Cultural Organization.

URSI *Union radio-scientifique internationale*/International Union of Radio Science.

VHF Very High Frequencies: 30-300 MHz.

VLF Very Low Frequencies: 10-30 kHz.

WARC World Administrative Radio Conference.

WARC-BS World Administrative Radio Conference for the Planning of the Broadcasting-Satellite Service in Frequency Bands 11.7-12.2 GHz (in Regions 2 and 3) and 11.7-12.5 GHz (in Region 1), 1977.

WARC-77 See WARC-BS, above.

WARC-ORB See Space WARC, above.

WARC-ST World Administrative Radio Conference for Space Telecommunications, 1971.

Bibliography

I. ITU DOCUMENTS

A. Plenipotentiary Conference Documents

International Telecommunication Convention, Madrid, 1932.
International Telecommunication Convention, Geneva, 1959.
International Telecommunication Convention, Montreux, 1965.
International Telecommunication Convention, Málaga-Torremolinos, 1973.
International Telecommunication Convention, Nairobi, 1982.

B. Radio Conference Documents

ITU Radio Regulations, 1959.
Final Acts of the Extraordinary Administrative Radio Conference to Allocate Frequency Bands for Space Radiocommunication Purposes, Geneva, 1963.
Final Acts of the World Administrative Radio Conference for Space Telecommunications, Geneva, 1971.
Final Acts of the World Administrative Radio Conference for the Planning of the Broadcasting-Satellite Service in Frequency Bands 11.7-12.2 GHz (in Regions 2 and 3) and 11.7-12.5 GHz (in Region 1).
Final Acts of the World Administrative Radio Conference, Geneva, 1979, Vols. I and II.
ITU Radio Regulations, 1982, Vols. I and II.
Final Acts of the World Administrative Radio Conference on the Use of the Geostationary-Satellite Orbit and the Planning of Space Services Utilizing It, Geneva, 1985.

World Administrative Radio Conference on the Use of the Geostationary-Satellite Orbit and the Planning of Space Services Utilizing It, First session, Geneva, 1985, *Report to the Second Session of the Conference.*

C. Other ITU Documents

Technical Bases for the World Administrative Radio Conference on the Use of the Geostationary-Satellite Orbit and the Planning of the Space Services Utilizing It (WARC-ORB (1)) *Report of the CCIR Conference Preparatory Meeting (CPM), Joint Meeting, Study Groups 1,2,4,5,7,8,9,10 and 11 Geneva, 25 June - 20 July 1984, Parts I and II.*

Regional Administrative Radio Conference 1983 for the Planning of the Broadcasting-Satellite Service in Region 2, *Report to the Administrations Prepared by the Panel of Experts of Countries of Region 2 with the Assistance of the IFRB, Geneva, March 1983.*

II. OTHER INTERNATIONAL DOCUMENTS

A. United Nations Documents

UN General Assembly Resolution 1348 (XIII). Question of the Peaceful Use of Outer Space, *13 December 1958.*

UN General Assembly Resolution 1472 (XIV). International Co-Operation in the Peaceful Uses of Outer Space, *12 December 1959.*

UN General Assembly Resolution 1721 (XVI). International Co-Operation in the Peaceful Uses of Outer Space, *20 December 1961.*

UN General Assembly Resolution 1802 (XVII). International Co-Operation in the Peaceful Uses of Outer Space, *14 December 1962.*

UN General Assembly Resolution 1962 (XVIII). Declaration of Legal Principles Governing the Activities of States in the Exploration and Use of Outer Space, *13 December 1963.*

UN General Assembly Resolution 1963 (XVIII). International Co-Operation in the Peaceful Uses of Outer Space, *13 December 1963.*

UN General Assembly Resolution 2453 (XXIII). International Co-Operation in the Peaceful Uses of Outer Space, *20 December 1968.*

Ad Hoc Committee on the Peaceful Uses of Outer Space. *Report. Official Records of the General Assembly, Fourteenth Session, 1959, Annexes,* (UN Doc. No. A/4141).

Ad Hoc Committee on the Peaceful Uses of Outer Space. *Report. Official Records of the General Assembly, Nineteenth Session, 1964-65. Annexes,* Vol. 2, (UN Doc. No. A/5785).

B. Multilateral Agreements on Space Activities

Treaty on Principles Governing the Activities of States in the Exploration and Use of Outer Space, Including the Moon and Other Celestial Bodies, signed at Washington, London and Moscow, *January 27, 1967,* 18 U.S.T. 2410, T.I.A.S. No. 6347, 610 U.N.T.S. 205 No. 8843.

Agreement Establishing Interim Arrangements for a Global Commercial Satellite System, signed *August 20, 1964,* 15 U.S.T. 1705, T.I.A.S. No. 5646.

Agreement Relating to the International Telecommuncations Satellite Organization, August 20, 1971, (entered into force *February 12, 1973*) 23 U.S.T. 3813, T.I.A.S. No. 7532.

Operating Agreement Relating to the International Telecommunications Satellite Organization, August 20, 1971, (entered into force *February 12, 1973*) 23 U.S.T. 3892, T.I.A.S. No. 7582.

III. US GOVERNMENT DOCUMENTS

Documents on International Aspects of the Exploration and Use of Outer Space, 1954-1962, *Staff Report, Senate Committee on Aeronautical and Space Sciences,* 88th Congress, 1st Session, Doc. No. 18, May 9, 1963.

U.S., Department of State, Office of International Communications Policy, *Report of the United States Delegation to the World Administrative Radio Conference for the Planning of the Broadcasting-Satellite Service in Frequency Bands 11.7-12.2 GHz (in Regions 2 and 3) and 11.7-12.5 GHz (in Region 1), Geneva, Switzerland, January 10-February 13, 1977,* TD Serial No. 80, Washington, DC.

U.S., Department of State, Office of International Communications Policy, *Report of the Chairman of the United States Delegation to the World Administrative Radio Conference of the International Telecommunication Union, Geneva, Switzerland, September 24-December 6, 1979,* TD Serial No. 116, Washington, DC.

U.S., Department of State, Office of International Communications Policy, *Report of the United States Delegation to the Plenipotentiary Conference of the International Telecommunication Union, Nairobi, Kenya, September 28-November 6, 1982,* Washington, DC.

U.S., Department of State, Office of International Communications Policy, *Report of the United States Delegation to the ITU World Adminstrative Radio Conference for the Mobile Services, Geneva, Switzerland, February 28-March 18, 1983,* TD Serial No. 123, Washington, DC.

U.S., Department of State, Office of International Communications Policy, *Report of the United States Delegation to the ITU Region 2 Administrative Radio Conference on the Broadcasting Satellite Service, Geneva, Switzerland, June 13-July 17, 1983,* Washington, DC.

U.S., Department of State, Office of International Communications Policy, *Report of the United States Delegation to the First Session of the ITU World Administrative Radio Conference on the Planning of the Geostationary Satellite Orbit and the Space Services Utilizing It, Geneva, Switzerland, August 8-September 16, 1985,* Washington, DC.

U.S., Congress of the United States. Office of Technology Assessment, *Radiofrequency Use and Management: Impacts of the World Administrative Radio Conference of 1979,* Washington, DC.

U.S., Federal Communications Commission, *First Report of the Advisory Committee for the ITU's World Administrative Radio Conference on the Use of the Geostationary-Satellite Orbit and the Planning of the Space Services Utilizing It, December 1983,* Washington, DC.

U.S., Federal Communications Commission, *Second Report of the Advisory Committee for the ITU World Administrative Radio Conference on the Use of the Geostationary-Satellite Orbit and the Planning of the Space Services Utilizing It, January 1985,* Washington, DC.

IV. BOOKS AND MONOGRAPHS

Codding, George A., Jr., *The International Telecommunication Union, An Experiment in International Cooperation,* Leiden: E.J. Brill, 1952. Republished in 1972 by Arno Press, New York.

Codding, George A., Jr., and Anthony M. Rutkowski, *The International Telecommunication Union in a Changing World,* Dedham, MA: Artech House, 1982.

DeFleur, M.L. and S. Ball-Rokeach, *Theories of Mass Communication,* (4th Ed.), New York: Longman, 1982.
Demac, Donna A., George A. Codding Jr., Heather E. Hudson and Ram S. Jakhu, *Access to Orbit: After the 1985 ITU Space WARC,* London: International Institute of Communications, 1986.
Feldman, Mildred L.B., *The Role of the United States in the International Telecommunication Union and Pre-ITU Conferences,* Baton Rouge, LA: Mildred L. Bos Feldman, 1976.
Gerbner, George and Marsha Siefert (eds.), *World Communications A Handbook,* New York & London: Longman, 1984.
Haley, Andrew G., *Space Law and Government,* New York: Appleton-Century-Crofts, 1963.
Hewitt, Paul G., *Conceptual Physics: A New Introduction to Your Environment,* Boston: Little, Brown and Company, 1977.
Intelsat Report 1984-1985, Washington, DC: Intelsat, 1985.
Jasentuliyana, Nandasiri and Roy S.K. Lee (eds.), *Manual on Space Law,* Vols. I and II, Dobbs Ferry, NY: Oceana Publications, 1979.
Leive, David M., *International Telecommunications and International Law: The Regulation of the Radio Spectrum,* Dobbs Ferry, NY: Oceana Publications, 1970.
Levin, Harvey J., *The Invisible Resource: Use and Regulation of the Radio Spectrum,* Baltimore: Johns Hopkins Press, 1971.
Martin, James, *Communications Satellite Systems,* Englewood Cliffs, NJ: Prentice-Hall, 1978.
McDougal, Myres S., Harold D. Lasswell and Ivan A. Vlasic, *Law and Public Order in Space,* New Haven: Yale University Press, 1963.
McLuhan, Marshall, *Understanding Media: The Extensions of Man,* New York: McGraw-Hill, 1964.
Meagher, Christine A., *Satellite Regulatory Compendium,* Potomac, MD: Phillips Publishing, 1985.
Nordenstreng, Kaarle and Herbert I. Schiller (eds.), *National Sovereignty and International Communication,* Norwood, NJ: Ablex, 1979.
Parry, A., *Russia's Rockets and Missiles,* New York: Doubleday, 1960.
Pfeiffer, John E., *The Emergence of Man,* New York: Harper & Row, 1969.
Pfeiffer, John E., *The Emergence of Society,* New York: McGraw-Hill, 1977.
Robinson, George S., and Harold M. White Jr., *Envoys of Mankind: A Declaration of First Principles for the Governance of Space Societies,* Washington, DC: Smithsonian Institution Press, 1986.
International Telecommunication Union, *From Semaphore to Satellite,*

Geneva: ITU, 1965.
Smith, Delbert D., *International Telecommunication Control: International Law and the Ordering of Satellite and Other Forms of International Broadcasting,* Leiden: A.W. Sijthoff, 1969.
Lay, S. Houston and Howard J. Taubenfeld, *The Law Relating to Activities of Man in Space,* Chicago: University of Chicago Press, 1970.
Toffler, Alvin, *The Third Wave,* New York: Bantam Books, 1980.
Wallenstein, Gerd D., *International Telecommunication Agreements,* Dobbs Ferry, NY: Oceana Publications, 1977.

V. PERIODICALS, CHAPTERS IN BOOKS, PROCEEDINGS

Busak, Jan, "Space Telecommunications at Present and in Future," *Proceedings of the Twenty-Second Colloquium on the Law of Outer Space, International Institute of Space Law of the International Astronautical Federation,* Munich, Germany, September 16-22, 1979, pp. 29-34.
Butler, Richard E., "Satellite Communications: Regulatory Framework and Applications for Development," *Space Communications and Broadcasting,* Vol. 3, No. 2, 1985, pp. 103-108.
Butler, Richard E., "World Administrative Radio Conference for Planning Broadcasting Satellite Service," *Journal of Space Law,* Vol. 5, Nos. 1 & 2, 1977, pp. 93-99.
Christol, Carl Q., "The International Telecommunication Union and the International Law of Outer Space," *Proceedings of the Twenty-Second Colloquium on the Law of Outer Space, International Institute of Space Law of the International Astronautical Federation,* Munich, Germany, September 16-22, 1979, pp. 35-51.
Clarke, Arthur C. "Extraterrestrial Relays, Can Rocket Stations Give World-Wide Coverage?" *Wireless World,* 1945.
"Coming to Consensus in Geneva," *Broadcasting,* July 18, 1983, pp. 24-26.
Colino, Richard R., "A Chronicle of Policy and Procedure: The Formulation of the Reagan Administration Policy on International Satellite Telecommunications," *Journal of Space Law,* Vol. 13, No. 2, 1985, pp. 103-156.
Colino, Richard R., "International Cooperation Between Communications Satellite Systems: An Overview of Current Practices and Future Prospects," *Journal of Space Law,* Vol. 5, Nos. 1 & 2, 1977, pp. 65-92.
Dalfen, Charles M., "Principles Governing Direct Satellite Broadcasting," in Jasentuliyana, Nandasiri, and Roy S.K. Lee

(eds.), *Manual on Space Law,* Vol. 1, 1979, pp. 283-302.

Dordick, Herbert S., "Transborder Data Flows: Competition and Regulation," in Gerbner, George and Marsha Siefert (eds.), *World Communications: A Handbook,* New York & London: Longman, 1984, pp. 210-216.

Doyle, Stephen E., "International Telecommunication Union WARC-85 Conference Preparatory Meeting," *Journal of Space Law,* Vol. 12, No. 2, 1984, pp. 174-182.

Freeman, Jennifer, "Direct Broadcast Developments and Directions: The National Sovereignty and Cultural Integrity Positions," *Proceedings of the Seventy-Fourth Annual Meeting, American Society of International Law,* Washington, DC, 1981, pp. 302-321.

Goedhuis, D., "Telecommunications by Satellite," *The International Law Association Report of the Fifty-Sixth Conference Held at New Delhi,* December 29, 1974-January 4, 1975, pp. 467-510.

Jasentuliyana, Nandasiri, "Regulations Governing Space Telecommmunication," in Jasentuliyana, Nandasiri, and Roy S.K. Lee (eds.), *Manual on Space Law,* Vol. 1, 1979, pp. 195-238.

Johnsen, Katherine, "U.S. Seeks Better Frequency Sharing," *Aviation Week and Space Technology,* July 23, 1979, pp. 65-67.

Kopal, Vladimir, "Evolution of the Main Principles of Space Law in the Institutional Framework of the United Nations," *Journal of Space Law,* Vol. 12, No. 1, 1984, pp. 12-25.

Kwen, Chen, "Pending Issues Before the Legal Subcommittee of the United Nations Committee on Peaceful Uses of Outer Space," *Journal of Space Law,* Vol. 5, Nos. 1 & 2, 1977, pp. 29-35.

Masmoudi, Mustapha, "The New World Information Order," in Gerbner, George and Marsha Siefert (eds.), *World Communications: A Handbook,* New York & London: Longman, 1984, pp. 14-27.

Mili, M., "World Administrative Radio Conference for the Planning of the Broadcasting-Satellite Service in Frequency Bands 11.7-12.2 GHz (In Regions 2 and 3) and 11.7-12.5 GHz (In Region 1)," *Proceedings of the Twentieth Colloquium on the Law of Outer Space, International Institute of Space Law of the International Astronautical Federation,* Prague, Czechoslovakia, September 26-October 1, 1977, pp. 346-367.

Ploman, Edward W., "Satellite Broadcasting, National Sovereignty and Free Flow of Information," in Nordenstreng, Kaarle and Herbert I. Schiller (eds.), *National Sovereignty and International Communication,* Norwood, NJ: Ablex, 1979, pp. 154-165.

"RARC-83: Cooperation Eliminates Surprises," *Broadcasting,* July 1983, pp. 22-27.

"Report of the Chairman of the Working Group on Direct Broadcast Satellites," *Journal of Space Law,* Vol. 8, No. 1, 1980, pp. 65-70.

Robinson, Glen O., "Regulating International Airwaves: The 1979 WARC," *Virginia Journal of International Law,* Vol. 21, No. 1, 1980, pp. 1-54.

Rutkowski, Anthony M., "Six Ad-Hoc Two: The Third World Speaks Its Mind," *Satellite Communications,* Vol. 4, March 1980, pp. 22-27.

Rutkowski, Anthony M., "Space WARC: The Stake of Developing Countries," *Space Policy,* Vol. 1, No. 2, 1985, pp. 240-243.

"Satellite Slots," *Fortune,* June 24, 1985, p. 88.

Smith, Milton J., "Space WARC 1985: The Quest for Equitable Access," *Boston University International Law Journal,* Vol. 3, 1985, pp. 229-255.

"Space WARC Reaches Consensus," *Broadcasting,* September 16, 1985, pp. 40-42.

"Space WARC Hits Obstacles in Geneva," *Broadcasting,* September 9, 1985, pp. 40-42.

Srirangan, T.V., "Equity in Orbit: Planned Use of a Unique Resource," paper presented to the 1984 Annual Conference of the IIC, Berlin, September 21-23, 1984.

Staple, Gregory C., "The New World Satellite Order: A Report from Geneva," *American Journal of International Law,* Vol. 80, No. 3, 1986, pp. 699-720.

"The Way It Was and Wasn't at WARC '85," *Broadcasting,* November 4, 1985, pp. 70-72.

"US Runs into Roadblocks Over DBS Plans," *Broadcasting,* July 4, 1983, p. 23.

"WARC is Over: U.S. 'Satisfied' says Robinson," *Broadcasting,* December 10, 1979, pp. 82-86.

"WARC '79: Curtain Going Up On Telecommunications Future," *Broadcasting,* September 17, 1979, pp. 35-45.

"WARC 1985: The Politics of Space," *Broadcasting,* September 23, 1985, pp. 56-58.

"Why the Sky Didn't Fall at WARC," *Broadcasting,* December 17, 1979, pp. 44-52.

"Will There Be Room on the Arc?" *Science,* Vol. 223, March 9, 1984, pp. 1043-1045.

Withers, David, "Results of WARC-ORB (1) Conference on Space Radio Services," *Space Communication and Broadcasting,* Vol. 3, No. 4, 1985, pp. 347-350.

Index

www.ingramcontent.com/pod-product-compliance
Lightning Source LLC
LaVergne TN
LVHW101458200926
848483LV00001B/157